T Level Animal Care and Management

ANIMAL MANAGEMENT CORE PATHWAY

Revision Guide

Gemma Hodgson
Carolyn Holehouse
Ben Lakin-Mason

eboru

While this book is designed to help and support teachers and learners throughout the course, the only official source of information about the qualification is the qualification specification, published by the awarding organisation. Teachers delivering this qualification should always refer to the specification for definitive information about all aspects of this qualification.

The questions in this book are designed to help learners develop their knowledge, skills and understanding. They are not assessment questions and they, along with the suggested marks and suggested answers, have not been seen, verified or approved by the awarding organisation.

Please refer to the awarding organisation for all matters relating to assessment and mark schemes.

Answers are available online. Please visit: www.eboru.com/TL-Answers

The publisher gratefully acknowledges the permission of copyright holders to reproduce copyright material.

Cover image: Kurit afshen/Shutterstock

Every effort has been made to trace copyright holders and to obtain their permission for the use of copyright material. The publisher will be glad to make arrangements with any copyright holder it has not been possible to contact.

First edition 2025. Impression 10 9 8 7 6 5 4 3 2 1

ISBN 978-1-917048-01-9

Whilst every effort has been made to ensure all information in this book is correct, the publisher shall not be liable for any loss of profit or any other commercial damages, including but not limited to special, incidental, consequential, personal, or other damages, due to any information or advice contained in this book.

If you do spot any errors in this book you can alert the publisher at: enquiries@eboru.com

Ordering Information

Special discounts are available for class set purchases by schools, colleges and others. For details, contact the publisher at: orders@eboru.com

Trade orders: copies of this book are available through the normal wholesalers. For any queries please contact: orders@eboru.com

www.eboru.com

Contents

Features in this book

Important Terms!
Key words are defined throughout the book

Remember!
Key points to take away as part of your revision

Recap Questions
Test your knowledge at the end of each sub-section

Practical Tips for Animal Care

- Use feed buckets with ergonomic handles.
- Employ rolling carts for heavy bedding or feed bags.
- Work with colleagues when lifting or restraining large animals.

Remember!
The "Individual Capability" principle is most relevant. Understanding personal physical limits and leveraging training ensures safe practices and reduces the likelihood of injury.

Important terms!
Capability: An individual's physical ability to handle objects safely.
Lifting Techniques: Correct methods to prevent injury, such as bending the knees and using the legs.
Assistance: Seeking help when a task is too difficult or dangerous to do alone.

Real-World Example: Principles of Safe Manual Handling in the Animal Management Sector

You're a new animal care assistant at a wildlife rescue centre, where you work closely with injured and orphaned animals. One day, you need to safely transport a small but injured rabbit from its recovery pen to the veterinarian for treatment. The rabbit has been recovering from a leg injury, and while it's calm, it's still important to handle it safely to prevent further stress or injury.

Applying Safe Manual Handling Principles

Assess the Risk

Before lifting or handling the rabbit, you take a moment to assess the situation. The rabbit is small, but injured, so you know that you need to minimise any jarring movements. The environment is clean, but there may be obstacles like food bowls or bedding that could trip you, so you plan your movements carefully.

Plan

Since the rabbit is small and fragile, you plan to use a soft towel to gently pick it up. This will provide both support and protection to prevent injury to the animal. You also ensure that the transport box for the rabbit is nearby and easy to access, so you won't have to hold the rabbit for longer than necessary.

Keep a Straight Back, Bend Knees

You carefully squat down with your back straight, bending at the knees, not the waist. This technique is key in avoiding strain on your back muscles. You scoop up the rabbit gently, supporting its body from underneath with the towel, ensuring that it feels secure in your arms.

Maintain a Firm Grip

You hold the towel around the rabbit's body securely but gently, avoiding any pressure on its injured leg. By maintaining a firm, balanced grip, you can safely carry the rabbit without the risk of dropping it or causing further injury.

Communicate with Others

On your way to the vet's office, you come across a colleague. You calmly inform them that you're transporting the rabbit and ask for a hand if you need it in case of any further challenges. Good communication is vital, especially when handling live animals, as they can be unpredictable.

Minimise the Distance

You take the most direct route to the vet's office, avoiding unnecessary steps or detours. This reduces the time the rabbit has to be handled, helping to avoid stress and injury.

Recap Questions

1. What are the three key pieces of legislation relevant to safe manual handling in the animal management sector?

2. Explain how the Manual Handling Operations Regulations 1992 (MHOR) impact the responsibilities of employers and employees when handling animals or equipment.

3. Why is risk assessment considered a key principle in ensuring safe manual handling in the animal management sector? Provide two examples of its application.

4. Define a contingency plan and explain its importance in the context of animal management facilities.

5. What are two key responsibilities of an employer under the Manual Handling Operations Regulations 1992 (MHOR) in the animal management sector?

6. List three consequences an employer might face if they fail to comply with manual handling legislation.

7. Describe one safe manual handling technique and explain why it is important in the animal management sector.

8. What are the five key factors to consider according to the HSE directive when performing manual handling tasks?

9. Why is it important to assess the working environment before lifting a load?

10. What should you do if a load is too heavy or awkward to lift on your own?

11. Why is it important to consider an individual's physical capability before attempting to lift or move objects in the animal management sector?

12. When is it necessary to summon assistance to move objects or loads? Name and explain two of the five factors that should be assessed when deciding whether help is needed.

13. How do the size and shape of objects impact manual handling and the need for assistance?

Practice Questions

1. Explain the importance of maintaining a clean and organised workspace in animal care. (4 marks)

2. Describe two key steps involved in conducting a risk assessment in an animal management setting. (4 marks)

3. Identify and explain two types of hazards commonly found in animal management facilities and the potential risks associated with each. (6 marks)

4. List three responsibilities of an employer in maintaining health and safety in an animal care environment. (6 marks)

5. Explain why it is essential to follow proper lifting techniques when handling large animals or heavy equipment in an animal care facility. (3 marks)

6. List and explain two legal responsibilities that animal care staff have in ensuring health and safety in the workplace. (4 marks)

7. Describe two environmental hazards that may arise in an outdoor animal enclosure and how to mitigate these risks. (5 marks)

8. Identify three potential risks when working with unfamiliar animals and provide one example of a control measure for each. (6 marks)

Real-world Examples
Shows how some of the theories and ideas apply in real animal settings

Practice Questions
Higher-level questions to test your knowledge and understanding

Photo credits

1.1 Hazards, risks and control measures associated with working in the animal management sector

Legislation and lone working

Lone working responsibilities

Lone working refers to situations where an employee works without close or direct supervision, which is common in animal management roles. This requires both employees and employers to adhere to specific responsibilities under health and safety legislation.

Employee Responsibilities:

1. **Follow Training and Guidelines**. Employees must follow the health and safety training provided, comply with safety protocols, and use personal protective equipment (PPE) appropriately.

2. **Risk Awareness and Reporting**. Lone workers should be vigilant about potential hazards and report any issues or risks immediately to their employer.

3. **Self-Monitoring and Communication**. In cases where immediate assistance isn't available, lone workers need to stay alert, monitor their own health, and ensure they

> **Important terms!**
>
> Lone Working: Working without immediate assistance or supervision.
>
> Risk Assessment: The process of identifying, evaluating, and managing risks in the workplace.

maintain scheduled communication with a supervisor or team member.

Employer Responsibilities:

1. **Risk Assessment and Hazard Management**. Employers must conduct thorough risk assessments to identify and manage potential risks associated with lone working, taking into account the specifics of animal handling.

2. **Implementing Safe Systems of Work**. Employers should implement safe systems, including procedures for regular check-ins and emergency contacts for lone workers.

3. **Health and Safety Policy**. A company health and safety policy should outline the protections and requirements for lone workers, ensuring both compliance and a clear approach to mitigate risks.

Relevant legislation for lone working

Health and Safety at Work Act 1974. This key legislation mandates that employers take reasonable steps to ensure the health, safety, and welfare of their employees, including those working alone.

Management of Health and Safety at Work Regulations 1999. These regulations expand on the Health and Safety at Work Act, requiring employers to conduct risk assessments, provide necessary training, and implement appropriate safety measures for lone workers.

> **Remember!**
> Understanding the Health and Safety at Work Act 1974 is crucial. This legislation serves as the foundation for health and safety practices across the sector, encompassing numerous other regulations and guidelines that ensure workplace safety.

How legislation and regulations impact safe working

The animal management sector must comply with multiple health, safety, and environmental regulations, ensuring the welfare of employees, animals, and the public. Here's an overview of key legislation:

Health and Safety at Work Act 1974 - The primary legislation governing workplace safety across sectors, this act obligates employers to protect their employees from risks to health and safety. It forms the basis for all other health and safety regulations, requiring risk assessments, training, and preventive measures in the workplace.

Reporting of Injuries, Diseases, and Dangerous Occurrences Regulations (RIDDOR) - RIDDOR mandates the reporting of serious workplace incidents, including injuries, diseases, and dangerous occurrences. In animal management, this could include reporting animal bites, zoonotic infections, or injuries caused by equipment.

Equality Act 2010 - This act ensures that all employees are treated fairly, prohibiting discrimination and promoting equal opportunities, including access to safety resources and risk assessments for workers with disabilities.

Provision and Use of Work Equipment Regulations 1998 (PUWER) - These regulations specify that all equipment used at work must be suitable, well-maintained, and safe to operate. Animal management facilities must ensure that equipment such as cages, feeding mechanisms, and cleaning tools are safe and regularly inspected.

Waste from Electrical and Electronic Equipment (WEEE) and PPE at Work Regulations - These regulations mandate the safe disposal of electrical waste and the provision of appropriate **PPE**. Animal care facilities must ensure safe disposal of electronic equipment and supply necessary PPE, such as gloves and goggles, to minimise risk.

Control of Substances Hazardous to Health Regulations (COSHH) - COSHH requires employers to control exposure to hazardous substances, which may include disinfectants, cleaning chemicals, and medications used in animal care. Proper storage, handling, and training for these substances are essential.

Manual Handling Operations Regulations 1992 - These regulations require employers to minimise risks related to manual handling, relevant in animal care due to the lifting of heavy objects or animals. Employers should provide training on safe handling techniques to avoid injury.

Health and Safety (First Aid) Regulations 1981 - This regulation requires employers to provide adequate first aid arrangements, ensuring access to first aid kits and trained personnel, which is particularly important in facilities where employees work with unpredictable animals.

Electricity at Work Regulations 1989 - These regulations address the safety of electrical installations and equipment. Animal management facilities must ensure that all electrical systems are safe, regularly inspected, and maintained to prevent electrical hazards.

Regulatory Reform (Fire Safety) Order 2005 - This law requires that workplaces have suitable fire safety measures, including fire alarms, evacuation procedures, and fire extinguishers. Regular fire drills are crucial in environments with animals to ensure safety for both staff and animals.

Control of Waste Regulations (England and Wales) 2012 - These regulations mandate that waste is disposed of in a manner that minimises risk to health and the environment. This includes proper disposal of animal waste, contaminated bedding, and chemicals used in animal care.

Environmental Protection Act 1990 - This act protects the environment from harm caused by workplace activities. Animal management facilities must follow guidelines to prevent environmental contamination, especially when disposing of animal waste or using hazardous chemicals.

Smoke-Free (Premises and Enforcement) Regulations 2006 - These regulations prohibit smoking in indoor workplaces to ensure a smoke-free environment, protecting employees and animals from secondhand smoke.

Health and Safety (Safety Signs and Signals) Regulations 1996 - This regulation mandates clear safety signage in workplaces. In animal management, signs should warn of potential hazards, such as animal bites, high-risk areas, and PPE requirements.

Types of hazards, risks, and control measures

Common hazards in the animal management sector

Animal care environments present multiple hazards that can pose threats to workers, animals, and the public. Key hazards include:

- **Lone Working**: Working alone can increase the risk of delayed emergency assistance. For example, if a worker is injured while alone, it may be difficult to call for help.

- **Equipment**: Machinery like feeders, heating systems, and power tools can present risks of injury, especially when not maintained or operated properly.

- **Uneven Ground and Slippery Surfaces**: Outdoor areas or enclosures often have uneven terrain or wet floors, which can lead to slips, trips, and falls.

- **Hazardous Materials**: Chemicals, pesticides, and cleaning agents used around animals can be harmful if not handled properly.

- **Weather**: Rain, snow, or high winds can make outdoor areas dangerous, affecting visibility and footing, increasing the risk of accidents.

- **Chemicals and Anaesthetic Gases**: Chemicals used for cleaning or medical gases pose risks of inhalation, poisoning, or skin contact hazards.

- **Unpredictable Animal Behaviour**: Animals can display aggressive or unexpected behaviours that may harm handlers, such as kicking, biting, or sudden movements.

- **Zoonotic Diseases and Biohazards**: Diseases like rabies or ringworm can be transmitted from animals to humans, especially in environments with diverse animal species.

- **Handling and Movement of Animals**: Moving large or agitated animals poses risks of physical injury (e.g., bites, scratches) to handlers.

- **Working at Heights**: Accessing areas like aviaries or cleaning high enclosures introduces the risk of falls.

- **Radiation Exposure**: In veterinary practices, exposure to radiation from X-ray machines can be harmful without protective measures.

- **Machinery and Equipment Contact**: Handling equipment and machinery improperly or without training increases the risk of cuts, bruises, or worse injuries.

> ### Important terms!
> **Hazard**: Any source that has the potential to cause harm or adverse health effects.
>
> **Risk**: The likelihood and severity of harm resulting from exposure to a hazard.
>
> **Control Measure**: Actions or equipment used to reduce or eliminate the risk associated with a hazard.
>
> **Zoonoses**: Diseases transmitted between animals and humans.
>
> **Biosecurity**: Procedures intended to protect humans and animals from biological hazards.

Risks in the animal management sector

Working in animal care can expose individuals to a range of risks that are specific to the hazards present. Notable risks include:

- **Zoonoses**: Risks associated with contracting diseases transmitted by animals.

- **Physical Injuries (Crushing, Kicking, Biting)**: Physical contact with animals may result in injuries, particularly from large animals or those with defensive behaviours.

- **Contamination or Asphyxiation**: Working with hazardous materials can lead to inhalation risks, contamination, or even poisoning.

- **Slips, Trips, and Falls**: Uneven or slippery surfaces pose significant risks, especially when handling animals or equipment.

- **Allergens**: Exposure to animal fur, dander, or environmental allergens can lead to respiratory issues or skin reactions.

- **Drowning**: Water features or pools in enclosures pose drowning risks, particularly when working alone.

Control measures to manage hazards and reduce risks

To maintain safety in the animal management sector, the implementation of various control measures is crucial. These measures include:

1. **Communication and Monitoring**

- **Agreed Contact Times**: Establish regular contact times with a supervisor, especially during lone work shifts.

- **Carrying a Mobile Phone/Radio**: Ensures quick access to help in emergencies.

2. **Risk Assessments and Site Awareness**

- **Risk Assessments**: Identifying hazards and assessing risks in the environment helps prioritise control measures.

- **Location and Return Awareness**: Workers should inform others of their location and estimated return time to ensure they can be located quickly if needed.

3. **Biosecurity and Personal Hygiene**

- **Biosecurity Measures**: Includes quarantine for new animals, controlling waste disposal, and reducing disease transmission risks.

- **Vaccinations**: Workers should be up-to-date on vaccinations to protect against zoonotic diseases.

- **Personal Hygiene**: Washing hands frequently and wearing clean PPE helps reduce exposure to pathogens.

4. **Personal Protective Equipment (PPE)**

- **PPE**: Items such as gloves, masks, and protective clothing reduce contact with harmful substances and protect against bites or scratches.

5. **Animal Handling and Restraint Techniques**

- **Animal Isolation**: Isolating sick or dangerous animals reduces the risk to humans and other animals.

- **Proper Handling Techniques**: Workers should use appropriate handling and restraint methods to prevent animal-related injuries.

6. **Access Safety and Working Schedules**

- **Safe Means of Access**: Ensures safe entry and exit to enclosures or elevated areas to reduce the risk of falls.

- **Working Schedules**: Planned schedules allow for sufficient breaks and adequate staffing, reducing fatigue and improving attention to safety.

7. **Health and Welfare Policies**
- **Lone Working Policy**: A policy that outlines safe practices and emergency protocols for employees working alone.
- **Employee Health and Occupational Welfare**: Ensuring workers are physically fit and informed about risks to handle the demands of the job.

> **Remember!**
> Understanding the importance of risk assessments is key. By identifying hazards, assessing potential risks, and applying control measures, animal management professionals create a safer work environment for both staff and animals.

Hazards when optimising animal welfare and their environment

In animal management, safeguarding human health and animal welfare requires a thorough understanding of **hazards**, **risks**, and **control measures**. Hazards in this field stem from interactions with animals, environmental conditions, and the use of tools and chemicals. The main objective is to manage these risks to ensure a safe and productive environment for both animal care professionals and the animals they tend to.

Hazards
- **Biological Hazards**: Animals can carry zoonotic diseases that are transmissible to humans, such as salmonella, rabies, and ringworm. Contact with animal waste, saliva, fur, or skin can transmit diseases.
- **Chemical Hazards**: Cleaning agents, disinfectants, medications, and pesticides used in animal settings pose chemical hazards. If these are not handled carefully, they can lead to poisoning, skin irritation, or respiratory issues if inhaled.
- **Physical Hazards**: Handling animals can lead to bites, scratches, and kicks. Equipment like cages, carriers, and grooming tools also pose

> **Important terms!**
> Ventilation: Management of airflow in animal environments to control temperature and contaminants.

risks, especially if not maintained properly or used correctly.
- **Environmental Hazards**: Poor ventilation, extreme temperatures, and excessive noise or light in animal habitats can cause stress in animals, increasing the likelihood of aggressive behaviours and human injury.

Risks associated with hazards
- **Disease Transmission**: Close interaction with animals and their waste increases the likelihood of infection from zoonotic diseases.
- **Chemical Exposure**: Chemicals used in cleaning or medication can cause burns, allergic reactions, respiratory issues, or poisoning if handlers come into contact with or inhale toxic fumes.
- **Injuries from Animal Interaction**: Physical risks from animal interactions include

potential injuries from animal bites, scratches, and kicks. Sudden animal movements or defensive behaviours can cause accidents and injuries to handlers.
- **Stress and Aggressive Behaviour**: Poor environmental conditions can elevate stress in animals, making them more prone to aggression. This increases the risk of injury to humans, who may face unpredictable animal behaviour.

Control measures
- **Personal Protective Equipment (PPE)**: Gloves, masks, aprons, and goggles reduce exposure to biological and chemical hazards.
- **Proper Handling Techniques**: Training in safe animal handling minimises the likelihood of physical injury, as confident and calm handling prevents animal stress and aggression.
- **Environmental Control**: Regular cleaning, good ventilation, equipment maintenance, and clear labelling of hazardous materials contribute to a safe environment for both humans and animals.

> **Remember!**
> Proper handling techniques and PPE reduces the risks of biological and physical hazards.

Risks when working with a variety of animals

Hazards and risks in animal management

When working in animal management, students and professionals encounter a range of hazards and risks, particularly when handling large and potentially dangerous animals. Awareness of these risks and implementing effective control measures are essential in ensuring both human and animal safety.

Associated risks with specific species

- **Large Zoo Mammals**: Species like big cats, elephants, and primates present risks due to their size, strength, and natural behaviours. They can inflict serious injury if not handled appropriately.

- **Large Birds of Prey**: These birds have powerful talons and beaks and can perceive humans as threats, especially during feeding or nesting. Mishandling can lead to severe lacerations.

- **Venomous Animals**: Reptiles, insects, and arachnids with venom can be dangerous to handle due to the risk of bites or stings, which may cause severe allergic reactions, toxicity, or even fatalities.

- **Dangerous Dogs**: Aggressive dogs or breeds with a tendency toward territorial behaviour may bite or scratch if they feel threatened, particularly if inadequately socialised or stressed.

- **Category 1 Species**: Certain species require special licences due to their threat level to human health and safety. This category includes animals that pose a significant risk if containment or control measures are compromised.

- **Farm Animals**: Large animals such as cattle or horses can accidentally injure handlers with kicks, bites, or by trampling, especially in enclosed spaces or during handling activities.

Risks in high-risk animal categories

- **Domestic Animals**: Even familiar pets like dogs and cats can pose risks if frightened or handled in ways they perceive as threatening.

- **Feral Animals**: These animals, while technically domestic, have reverted to a wild state and can behave unpredictably, increasing the risk of injury.

- **Wild Animals (both free-roaming and captive)**: Wild animals, whether in their natural habitat or captivity, retain instinctual defence mechanisms that make handling dangerous.

- **Wild Captive Animals**: Animals in captivity that were originally wild may pose behavioural unpredictability and increased aggression when under stress.

High-risk situations

- **Direct Contact**: Physical interaction with high-risk animals is often necessary but increases the chance of injury or zoonotic disease transmission. Proper protective equipment, awareness of animal behaviour, and adherence to handling protocols are essential control measures to reduce the risk of injury.

Key control measures

- **Risk Assessments**: Conduct detailed assessments to understand the specific risks associated with each species and scenario.

- **Protective Equipment**: Use of gloves, face shields, and appropriate tools helps prevent injury and limit disease exposure.

- **Training and Protocols**: Ensure staff are well-trained in species-specific handling techniques and emergency procedures.

- **Biosecurity Practices**: Implement hygiene protocols to prevent zoonotic diseases and maintain safe environments for both humans and animals.

> **Remember!**
> Understanding the importance of control measures in mitigating hazards is crucial to ensure safety when working with high-risk animals and in high-risk environments.

Personal Protective Equipment (PPE)

Purpose of control measures

In animal management, control measures and precautions are critical for **managing and minimising risks** associated with handling animals, equipment, and chemicals. These measures help protect staff, animals, and visitors by reducing the likelihood of accidents and injuries, ensuring a safer environment. Understanding and applying control measures can prevent illnesses, injuries, and potential fatalities, while also promoting best practices and professionalism in the animal management sector.

> **Important terms!**
>
> Barrier Cream: A lotion that creates a protective layer on the skin.
>
> Cross-Contamination: Transfer of contaminants from one object, animal, or person to another.
>
> Respirator: A device designed to protect the wearer from inhaling hazardous particles.
>
> Steel-Toed Boots: Footwear reinforced with metal to protect feet from heavy objects.
>
> Ear Defenders: Equipment used to protect hearing from loud noises.

Personal protective equipment (ppe)

The correct use of PPE is essential in reducing exposure to risks, especially when working with animals or hazardous materials. Each type of PPE serves a specific purpose:

- **Eye Protection**: Protects against particles, chemicals, or other substances that may damage the eyes.

- **Masks** (Disposable, Respirator, Specialist Respiratory): Protects the respiratory system from harmful particles, dust, and biohazards. Respirators are essential in environments with airborne pathogens.

- Barrier Cream: Provides a protective layer on the skin to prevent irritation or allergic reactions caused by exposure to chemicals or animal substances.

- **Gloves** (Disposable, Gauntlets, Working): Protect hands from bites, scratches, and exposure to chemicals. Gauntlets offer additional arm protection, useful in animal handling.

- **Aprons**: Shields clothing and skin from liquids, contaminants, and dirt, reducing cross-contamination between different animals.

- **Overalls**: Offers full-body protection against animal dirt and pathogens, keeping regular clothing clean and preventing cross-contamination.

- **Appropriate Footwear**: Footwear such as steel-toed boots protects against crush injuries, while slip-resistant soles prevent falls in wet or slippery areas.

- Ear Defenders: Protects hearing from loud or persistent noises, especially in areas with barking dogs or industrial equipment.

- **Protective Headgear**: Shields the head from impact injuries, particularly in areas where large animals are handled.

> **Remember!**
>
> Understanding the purpose and correct usage of PPE is essential, particularly when completing a risk assessment. It is important to understand the benefits and limitations of different PPE methods and the risks they can reduce.

Dynamic risk assessing in different environments

Working in animal management requires **identifying and mitigating potential hazards** to ensure safety for animals, staff, and the public. One of the most crucial skills for professionals in this sector is **ongoing dynamic risk assessment**. This is a continuous process of assessing and responding to risks as they change in real-time, especially in varied animal environments. These environments often pose unique risks due to factors such as location (indoor vs. outdoor), conditions (weather), and interaction (public exposure).

- **Indoor and Outdoor Environments.** Indoor environments, like animal shelters and labs, require assessments of ventilation, cleanliness, and potential contamination risks. Outdoor environments, including farms, zoos, and natural habitats, present different risks, such as uneven terrain and exposure to unpredictable weather. Each setting needs tailored risk assessments to address location-specific hazards.

- **Field Work Hazards.** Field work in animal management can include collecting samples,

Pesticides used on farms present a hazard

> **Important terms!**
>
> Dynamic Risk Assessment: Continuous evaluation of changing risks in real-time.
>
> Contamination: The presence of harmful substances that may affect animal health and safety.
>
> Field Work: On-site animal management work that requires additional hazard considerations.
>
> Public Interaction: Engagements with the public, requiring specific safety protocols.
>
> Environmental Considerations: Factors in the environment that may affect risk levels.

observing wildlife, or rehabilitating animals in their natural habitats. Fieldwork introduces unique risks, such as difficult access to emergency services, limited shelter, and exposure to wild animals. In these settings, it's essential to conduct thorough pre-planning and make use of proper equipment to handle the varying conditions.

- **Weather Conditions.** Weather plays a significant role in animal management risks, particularly outdoors. Rain, snow, extreme heat, and strong winds can affect both animal and human safety. Weather can also impact animal behaviour, creating additional hazards. For example, extreme heat may cause dehydration in both humans and animals, while rain can make surfaces slippery and challenging to navigate.

- **Public Interactions.** Animal management often involves direct public engagement, such as in petting zoos, animal-assisted therapy, or educational tours. Public interactions require constant vigilance as unfamiliar individuals may inadvertently pose risks, either by not following safety protocols or through unintentional stress or harm to animals. Risk assessments should ensure the public is briefed on safety practices and closely monitored.

- **Environmental Considerations (Contamination).** Contamination risks include biological (disease transmission), chemical (pesticides), and physical hazards (debris, improper waste disposal). Implementing hygiene protocols, appropriate disposal systems, and frequent monitoring can mitigate these risks, protecting both the environment and those within it.

Hierarchy of control

In the animal management sector, understanding how to identify and manage risks is essential. Proper risk management uses a structured approach called the hierarchy of control measures, which minimises these risks by addressing them in order of effectiveness.

Hierarchy of Control Measures

- Elimination – The most effective way to reduce risk is to eliminate the hazard. This might involve redesigning a job to remove the risk, such as changing a work task to avoid animal contact or automating processes where possible.

- Substitution – If a hazard cannot be eliminated, consider replacing the process or material with something less dangerous. For example, use safer equipment or non-toxic cleaning agents to reduce the risk associated with handling chemicals.

- Engineering Controls – When elimination and substitution are not feasible, engineering controls help to isolate the hazard. This can include using physical barriers, such

> **Important terms!**
>
> Hierarchy of Control Measures: Stepwise approach to minimise risks.
>
> Elimination: Removing the hazard.
>
> Substitution: Replacing with a safer alternative.
>
> Engineering Controls: Physical separation of hazards.
>
> Administrative Controls: Policies and procedures to ensure safety.

as guards on machinery, fenced animal enclosures, or designated walkways. These methods prevent direct exposure to dangers and keep workers and animals separate where necessary.

- Administrative Controls – These are the policies and procedures developed to ensure safe working practices. For instance, proper training, certification requirements, clear signage, and emergency response protocols are all essential. Providing training in animal behaviour can help workers anticipate risks, while regular safety checks ensure consistent compliance with procedures.

- **Personal Protective Equipment (PPE)** – As the last line of defence, PPE is used when other control measures cannot fully eliminate the risk. Workers may wear gloves, helmets, goggles, or reinforced clothing to protect themselves from hazards in the animal care environment. PPE is crucial but should not be relied on to reduce risk by itself.

> **Remember!**
> The hierarchy of control shows that PPE should be the last resort. Whilst PPE is important, it is at the bottom of the pyramid of hierarchy. This is because it is the only method that does not address the risk itself, instead providing a protective layer to the individual.

Real world example: Working with Farm Animals

Imagine you're a new worker at a farm, taking care of cattle. Your job includes feeding the cows, cleaning their pens, and ensuring they're healthy. However, working with animals comes with various hazards, risks, and control measures to keep everyone safe.

1. Hazard: Physical Injuries from Large Animals

Risk: Cattle are large and powerful animals. They may become agitated or frightened and could potentially push or kick someone. This could result in bruises, broken bones, or other serious injuries.

Control Measure: To reduce the risk of injury, the farm provides you with personal protective equipment (PPE), such as gloves, boots, and a high-visibility vest. You're also trained to approach cattle calmly and safely. Additionally, there are designated paths and safe zones where you should stand while feeding or interacting with the animals, to avoid being in their direct path.

2. Hazard: Zoonotic Diseases (Diseases Transferred from Animals to Humans)

Risk: Some farm animals, including cattle, can carry diseases like E. coli or Salmonella, which can be transmitted to humans through direct contact with animals or contaminated surfaces.

Control Measure: The farm implements strict hygiene protocols. You must wash your hands thoroughly after handling animals, using disinfectant stations, and wear gloves when cleaning pens. The farm also schedules regular health checks for the animals and provides vaccinations, reducing the likelihood of disease spread.

3. Hazard: Slips, Trips, and Falls

Risk: The barn or feeding areas can become slippery, especially during the winter or after heavy rainfall. Mud, manure, or wet surfaces can cause workers to slip, potentially resulting in sprained ankles or worse.

Control Measure: The farm ensures that walkways are kept clean and dry, and anti-slip mats are placed in high-risk areas. Additionally, you are provided with sturdy, non-slip footwear to prevent falls.

4. Hazard: Handling Hazardous Chemicals (e.g., Cleaning Agents, Pesticides)

Risk: The farm may use chemicals to clean the pens or treat the animals for parasites. Exposure to these chemicals could lead to respiratory issues, skin irritation, or poisoning if not handled properly.

Control Measure: You receive training on how to safely use and store chemicals, including wearing appropriate PPE like masks and goggles. The farm also uses safer, environmentally friendly chemicals whenever possible and ensures that all chemicals are clearly labelled with hazard warnings.

5. Hazard: Stress and Fatigue

Risk: Working long hours in challenging conditions, such as during calving season, can lead to physical and mental fatigue. This can affect concentration and increase the likelihood of accidents or mistakes.

Control Measure: The farm enforces strict work-hour regulations, ensuring you have regular breaks and adequate rest. Supervisors are also trained to recognise signs of fatigue and stress in workers and provide support when needed.

Recap Questions

1. Explain the main responsibilities of employers and employees in relation to lone working in the animal management sector under the Health and Safety at Work Act 1974.

2. Identify and briefly describe three specific regulations other than the Health and Safety at Work Act 1974 that impact safe working practices in the animal management sector.

3. Why is conducting a risk assessment important in the animal management sector, and which regulation specifically requires employers to carry out risk assessments for their employees)?

4. Define "biohazard" and give an example.

5. What risk does zoonotic disease pose to humans?

6. List two control measures for handling hazardous materials.

7. What is meant by "animal welfare"?

8. Name a common hazard when optimising animal habitats.

9. Why is training important in ensuring welfare standards?

10. List three types of animals classified as high-risk due to their potential for causing serious injury.

11. What is one control measure that should be used to reduce the risks associated with direct contact with high-risk animals?

12. Define the term "hazard" in the context of animal management.

13. What is a respirator used for?

14. Why is PPE important in animal care?

15. List two types of PPE and their uses.

16. What is a dynamic risk assessment?

17. Why is dynamic risk assessment important in animal fieldwork?

18. Name an environmental factor that affects risk.

19. What does "elimination" mean in risk control?

20. List two methods in the hierarchy of control.

21. Why is PPE the last resort in hazard management?

1.2 Procedures and contingency and emergency plans to follow when dealing with emergency situations

Emergency planning and situations

Effective emergency procedures and contingency plans are key to ensure the safety of everyone.

Plans and considerations

- Personnel: Ensure all staff members are trained for emergency responses, including fire evacuation, first aid, and safe animal handling. Regular drills and clear communication protocols should be established.

- Animal Welfare: Plans must prioritise the welfare of animals in emergencies. This includes ensuring safe evacuation, temporary housing, and provisions for feeding, watering, and medical needs.

- **Environment**: Emergency planning must minimise environmental risks, including preventing contamination or damage from substances like chemicals, and managing waste or debris that could harm animal habitats.

Types of emergency situations:

- **Fire**: Establish clear evacuation routes for both personnel and animals. Install fire alarms, sprinklers, and keep fire fighting equipment easily accessible.

- **Medical Emergencies**: Have a designated first aid area and trained first aid personnel available. Maintain a stock of essential medical supplies for both humans and animals.

- **Animal Escape**: Prepare containment protocols for potential escapes, including recapture equipment and designated safe areas to move escaped animals.

- **Animal Injury or Attack**: Staff should be trained to respond calmly and promptly to injuries or aggressive behaviour, with clear procedures for isolation and treatment.

- **Power Cut**: Power failure may disrupt lighting, heating, and essential equipment. Backup generators and manual tools should be in place to ensure basic operations can continue.

- Natural Disasters **and Extreme Weather**: Assess risks of local weather patterns and prepare accordingly, e.g., flood barriers, sandbags for flooding, and stable shelters for animals during extreme cold.

- **Disease Outbreaks**: Create protocols for isolation, disinfection, and restricted movement within facilities to control outbreaks.

- **Contact with electricity**, Gas Leak, Water Issues: Identify emergency shutdowns and procedures to prevent injuries or contamination.

Important terms!

Personnel: Staff responsible for animal care and facility safety.

Animal Welfare: The health, safety, and comfort of animals in care.

Evacuation: Organised movement of animals and personnel during emergencies.

Containment: Strategies to safely secure animals in emergency scenarios.

Isolation: Separating affected animals during disease outbreaks.

Natural Disaster: Severe environmental events affecting facility operations.

Contingency Plan: Pre-arranged procedures for responding to emergencies.

Remember!

Maintaining calm and efficient emergency responses for animals is central to effective animal management.

- **Unauthorised Public Access or Activist Activity**: Security measures, such as surveillance and restricted access, can reduce risks from unauthorised individuals or activist disturbances.

- **Staff Shortage and Equipment Failure**: Develop staff coverage plans and regular equipment maintenance checks to prevent disruptions in care and safety.

Legislation and reporting incidents

Key legislation impacting emergency planning

- **Health and Safety at Work Act 1974**: This act mandates employers to create a safe working environment, directly influencing emergency planning by requiring risk assessments and clear, accessible safety procedures.

- **Control of Substances Hazardous to Health (COSHH) 2002**: COSHH enforces controls over handling hazardous substances, which are essential in animal care environments where chemicals or animal-borne pathogens are present. Emergency plans must address containment and first-aid protocols for exposure incidents.

- Reporting of Injuries, Diseases and Dangerous Occurrences Regulations **(RIDDOR) 2013:** RIDDOR obligates employers to report specific injuries and dangerous incidents, ensuring transparency and learning from past incidents. Non-compliance can lead to legal repercussions, damaged reputation, and lowered staff morale.

- **Zoonoses Order 1989**: This legislation controls zoonotic diseases (diseases that can spread from animals to humans). Emergency plans must include isolation protocols for infected animals to prevent cross-contamination and protect staff.

- **Regulatory Reform (Fire Safety) Order 2005**: This order requires that fire safety measures, such as alarm systems and evacuation plans, are implemented and maintained. Animal management facilities must have specific fire procedures to account for safe animal evacuation.

- **Movement of Animals (England) Order 2006**: This order regulates animal transport, impacting emergency procedures involving evacuations and disease control, particularly in scenarios involving suspected infectious diseases.

- **Firearms Act 1968**: This act regulates firearm use. This is relevant to some animal

management facilities where humane euthanasia may be needed in emergencies.

- **Zoo Licensing Act 1981**: Requires zoos to have emergency plans that cover animal escape and public safety, with staff trained to manage potential high-risk situations.

Importance of Reporting and Compliance

Remember! Accurate and timely accident and incident reporting, as per RIDDOR, is essential. Reporting helps in identifying recurring risks, improving safety protocols, and fulfilling legal obligations. Failure to comply with RIDDOR and other regulations can lead to serious consequences, including legal action from the Health and Safety Executive (HSE), financial penalties, and reputational damage, which can impact staff morale and public trust.

- -

Emergency risks and control measures

In the animal management sector, emergency situations can pose **serious health and safety risks** to both humans and animals. Having **well-defined procedures and contingency plans** ensures that any risks are minimised and dealt with effectively.

Types of emergency situations

- **Fire** – Fires can occur due to faulty electrical equipment, improper storage of flammable materials, or accidental ignition. In an animal facility, fire alarms, sprinklers, and accessible extinguishers are essential. Staff should know evacuation routes and fire assembly points for both animals and people.

- **Animal Escapes** – Animals that escape from enclosures pose risks to the staff, the public, and other animals. Escape procedures should include protocols for safely containing and recapturing animals, such as using secure doors, fences, or enclosures, and having tranquillisers on hand when necessary.

- **Disease Outbreaks** – Contagious diseases, such as avian flu or rabies, can spread rapidly among animals and potentially affect humans. Quarantine protocols, regular

> **Important terms!**
>
> Quarantine: Isolating animals to prevent the spread of disease.
>
> Evacuation Routes: Predetermined paths for safe evacuation during emergencies.

vaccinations, and sanitary measures are necessary control strategies. Quick isolation of affected animals is crucial to prevent disease spread.

- **Natural Disasters** – Floods, storms, and other natural events can impact animal facilities. Emergency plans should outline evacuation procedures, backup power sources, and emergency supplies for sustained care.

- **Chemical Spills** – Chemical hazards (e.g., cleaning agents or medical supplies) can be dangerous if mishandled. Training on proper storage, handling, and immediate cleanup processes is necessary, along with protective equipment for staff.

Control measures

- **Risk Assessments** – Regular assessments identify potential hazards and help staff prepare.

- **Clear Emergency Protocols** – Procedures should be documented and visible to all staff, outlining roles, contact numbers, and necessary actions.

- **Regular Training** – Frequent drills and training keep staff prepared and familiar with emergency responses.

- **Safety Equipment** – Fire extinguishers, first aid kits, protective gear, and barriers should be available and maintained.

- **Communication Systems** – An efficient system for alerting all staff members, such as alarms or radios, is essential for fast responses.

> ### Remember!
> The importance of regular training and drills is key to ensuring staff are prepared for any emergency, as it allows them to respond effectively under pressure.

Emergency planning

In the field of animal care and management, **preparing for and responding effectively to emergencies** is critical. Emergencies can arise from **accidents**, **animal health crises**, or **environmental hazards**. Understanding the correct procedures for handling these situations, along with the importance of having well-developed contingency plans, is essential for maintaining safety, animal welfare, and professional standards.

> ### Important terms!
> Emergency Action Plan (EAP): A structured plan outlining roles and procedures during emergencies.
>
> Contingency Plan: A proactive plan detailing steps to handle potential emergency scenarios.
>
> First Aid: Initial care provided in response to injury or illness, either human or animal.
>
> Evacuation Protocol: Procedure for safely moving individuals and animals out of a danger zone.
>
> Incident Communicator: Role responsible for maintaining communication during emergencies.

Correct procedures in emergency situations

Communication in Emergencies

Effective communication is crucial during an emergency to ensure that everyone involved is informed and can respond appropriately. Communication must be:

- **Clear and direct**: Use concise language and avoid technical jargon.

- **Timely**: Inform all relevant staff immediately to avoid delays.

- **Documented**: Record all actions and decisions for follow-up and legal compliance.

For example, if an animal escapes, the first step is to communicate the risk to others on-site and engage any relevant emergency contacts, if needed.

Ensuring Safety of Self and Others

In emergencies, the **first priority is always the safety of humans before animals**. Follow these steps:

- **Evaluate the area**: Check for immediate dangers such as fire, hazardous chemicals, or unstable structures.

- **Remove or protect individuals at risk**: Move them to a safe location if possible.

- **Use personal protective equipment (PPE)**: Ensure PPE is worn to prevent exposure to harmful substances or aggressive animals.

Protocols help us to remain calm and focused on the protocol, which helps protect all involved and prevents escalation.

Delegation of Roles in an Emergency

An emergency action plan (EAP) should outline the roles and responsibilities of each team member. Assign roles based on individual skills and experience, such as:

- **First aid responder**: Provides immediate first aid to any injured person or animal.

- **Animal handler**: Trained member of staff to safely restrain or guide animals.

- Incident communicator: Person responsible for updating management, contacting external support, and maintaining documentation.

Role delegation reduces confusion, speeds up response times, and allows specialised personnel to manage specific tasks effectively.

Importance of contingency and emergency plans

Key Reasons for Contingency Plans

Effective contingency plans **anticipate potential emergencies** and **outline clear steps** for addressing them, helping to:

- **Minimise harm or injury**: Protecting both people and animals from preventable harm.
- **Safeguard animal welfare**: Ensuring that animals receive timely care and support during emergencies.
- **Avoid legal consequences**: Non-compliance with safety and animal welfare laws can lead to legal action.

- **Preserve reputation**: Prompt and responsible emergency management reflects positively on the facility.
- **Boost staff morale**: Staff members who feel prepared are more confident and committed to their roles.

A **lack of emergency plans can have severe consequences**, potentially leading to loss of credibility, financial penalties, and a demoralised workforce. For instance, if a facility lacks a fire response plan, a sudden fire could endanger lives and severely damage the business.

Staff training requirements

Animal and Human First Aid

First aid training equips staff to respond effectively to medical emergencies. This includes:

- **Human first aid**: Basic skills like CPR, wound care, and handling fractures can save lives.
- **Animal first aid**: Techniques like restraining injured animals, controlling bleeding, and assessing vital signs are essential for animal welfare.

Fire Safety

All staff must be trained in fire safety procedures, including:

- Evacuation protocols: Knowing the correct escape routes and assembly points.
- Use of fire extinguishers: Understanding the types and uses of different extinguishers.
- **Animal evacuation techniques**: Safely guiding animals to secure locations, especially if they are panicked.

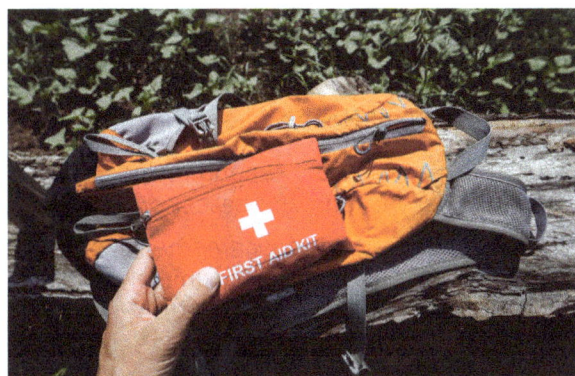

> ### Remember!
> An EAP guides each staff member in their role, ensures communication flows smoothly, and maintains safety for all.

Training should be repeated regularly to ensure all staff remember essential protocols, minimising risks in an actual fire.

Consequences of inadequate emergency plans

When contingency and emergency plans are **absent or poorly implemented**, the **consequences can be severe**. The impact includes:

- **Harm or injury**: Individuals may suffer due to delayed or unorganised response efforts.
- **Compromised animal welfare**: Animals may be left in vulnerable situations, leading to suffering or even death.
- **Legal repercussions**: Non-compliance with safety regulations can lead to fines, lawsuits, and loss of licensing.

- **Damage to reputation**: Poor emergency response can damage the credibility and public trust in an organisation.
- **Reduced staff morale**: Workers feel unprepared and anxious, leading to lower job satisfaction and potentially high turnover rates.

For example, if an animal care facility does not provide fire safety training, staff may be ill-equipped to respond, risking injury to both staff and animals and leading to costly legal repercussions.

Stakeholders and other organisations

Key organisations to notify in emergency situations

When an emergency occurs in the animal management sector, various organisations must be informed depending on the nature of the incident. Below are the primary agencies:

Department for Environment, Food and Rural Affairs (DEFRA)

DEFRA oversees policies related to animal health and welfare, including response protocols for zoonotic disease outbreaks or environmental hazards impacting animals.

Situations requiring DEFRA notification include major animal disease outbreaks, contamination risks, or environmental disasters affecting animal welfare.

Animal and Plant Health Agency (APHA)

The APHA, a part of DEFRA, handles specific incidents involving animal diseases or pests.

APHA must be contacted for emergencies such as suspected cases of notifiable animal diseases

> **Important terms!**
>
> DEFRA - Government department responsible for environmental protection, food production, and animal health and welfare.
>
> APHA - Agency under DEFRA managing animal diseases, biosecurity, and imports/exports.
>
> HSE - UK organisation ensuring workplace health, safety, and welfare compliance.
>
> Zoonotic Diseases - Diseases transmitted from animals to humans.

(e.g., avian influenza or foot-and-mouth disease).

They also manage animal import/export issues, biosecurity risks, and disease surveillance.

Health and Safety Executive (HSE)

The HSE ensures compliance with workplace safety laws, including those in animal care environments.

The HSE must be notified when an emergency poses risks to workers, such as accidents involving dangerous animals, hazardous substances, or structural failures at facilities.

Dangers to the public and services to notify

Emergencies in the animal management sector can also endanger the public. Identifying and managing these risks promptly is essential.

Key Dangers to the Public

Animal Escapes: Large or dangerous animals (e.g., exotic species, livestock) can cause injuries or traffic accidents.

Zoonotic Diseases: Diseases that can spread from animals to humans, such as rabies or leptospirosis, pose serious public health risks.

Hazardous Conditions: Broken enclosures, contaminated water supplies, or fire hazards in animal care facilities.

Services to Notify

Emergency Services (Police, Fire, Ambulance):

- Police: For public safety issues, such as animal escapes causing traffic hazards.
- Fire Brigade: For structural fires or flooding at

animal care facilities.

- Ambulance Services: When injuries occur due to animal attacks or hazardous environments.

Local Authorities. Local councils manage community safety and may provide resources during emergencies, such as temporary shelters or road closures.

Power Companies. Power companies should be notified if electrical hazards arise, such as downed lines affecting animal enclosures or veterinary equipment.

Health and Safety Executive (HSE). The HSE must be informed when an emergency involves potential workplace hazards that could affect staff or public safety, such as exposure to hazardous chemicals used in animal care facilities.

Procedures to follow in an emergency

Initial Assessment

- Evaluate the situation quickly and determine the nature and scale of the emergency.
- Identify immediate threats to life, property, or the environment.

Communication Plan

- Notify appropriate organisations and services based on the emergency type.
- Maintain clear records of communication, including the time, date, and contact details.

Containment and Control

Implement measures to control the situation, such as securing animals, isolating affected areas, or administering first aid to injured persons or animals.

Evacuation Plans

If necessary, evacuate animals and personnel following predetermined procedures.

Ensure transport vehicles and temporary shelters are ready and accessible.

Documentation and Follow-Up

- Record all actions taken during the emergency.
- Conduct a review of the incident to identify areas for improvement in future contingency plans.

> **Remember!**
> Focus on learning the roles of DEFRA, APHA, and HSE, along with recognising dangers to the public and the appropriate services to notify.

Animal provider emergencies

Emergency situations in animal management vary significantly across facilities, requiring **bespoke procedures** and contingency planning. Understanding these differences enables the swift and effective response necessary to protect animals, staff, and the public.

> **Important terms!**
> Isolation Procedure: Method for separating affected animals to prevent disease spread.

Emergency situations in different animal providers

Boarding Establishments: Emergencies may include animal escapes, severe weather, or infectious disease outbreaks. Plans should focus on evacuation protocols, isolation procedures, and communication with owners.

Veterinary Practices: Emergencies might involve fire, flooding, or medical equipment failure. Plans must prioritise staff and client safety, securing animals, and protecting sensitive medication or equipment.

Pet Shops: Fire hazards, power outages affecting heating or lighting, or aggressive animal behaviour are common risks. Contingency measures should include fire safety drills, backup power solutions, and protocols for handling distressed animals.

Wildlife/Farm Parks: Emergencies could include wild animal escapes, zoonotic disease outbreaks, or visitor injuries. Strategies include containment protocols, public evacuation routes, and liaison with emergency services.

Sanctuaries/Rescue/Rehabilitation Centres: Risks may involve natural disasters, resource shortages, or volunteer injuries. Contingency plans should cover evacuation, alternative housing arrangements, and medical assistance.

Private Collections: These may face emergencies such as exotic animal escapes or toxic exposure. Planning must emphasise securing enclosures and emergency responder coordination.

Breeders: Risks include birthing complications or disease spread. Protocols should address animal isolation, veterinary contact, and owner notifications.

Aquariums: Emergencies might entail water quality issues, filtration failure, or public safety incidents. Response plans should prioritise aquatic life support systems and crowd control.

Zoos: Major risks include large animal escapes, public injuries, and habitat damage. Key measures include enclosure security checks, public evacuation procedures, and inter-agency collaboration.

Grooming Establishments: Potential emergencies include animal bites or chemical spills. Plans should focus on first aid, equipment safety, and owner communication.

Pet Sitters: Emergencies may involve lost pets, sudden illness, or environmental hazards. Protocols should detail client notification and veterinary contact procedures.

Effective emergency management in animal care demands tailored planning based on the specific risks and operational goals of each facility type. By understanding these differences, animal care professionals can ensure welfare and safety standards are consistently upheld.

Real-World Example: Emergency Plan at an Animal Shelter

Imagine you're working at a large animal shelter that houses dozens of dogs, cats, and small animals. One afternoon, there is a sudden fire alarm, signalling a potential fire in the building. Here's how the procedures and contingency plans would unfold:

Emergency Situation: A fire has broken out in the shelter's kitchen area, and the building must be evacuated immediately to ensure the safety of both animals and staff.

Step-by-step procedures and contingency plans

1. Sound the Alarm

Procedure: The fire alarm is immediately activated to alert everyone inside the shelter.

Contingency Plan: If the alarm doesn't work, a staff member would use a loudspeaker or go to each room to alert everyone of the emergency.

2. Evacuation of People

Procedure: All staff and volunteers are trained to guide people and animals calmly to designated safe exit points.

Contingency Plan: If the main exit is blocked, staff have been trained on secondary evacuation routes, which lead to a safe outdoor area.

3. Evacuation of Animals

Procedure: Each animal has a pre-assigned handler or group of handlers who are familiar with how to quickly remove them from their enclosures in a safe and calm manner.

- Dogs are placed on leashes and walked to the exits.
- Cats are placed in secure carriers.
- Small animals are moved in their cages or carriers.

Contingency Plan: In case the shelter's usual animal transport equipment (like carriers) is damaged or inaccessible, staff have a plan to use alternatives such as crates or blankets to safely carry animals out.

4. Communication During the Emergency

Procedure: The shelter has a designated emergency contact person who communicates with the fire department and any nearby veterinary clinics to ensure the animals' well-being.

Contingency Plan: If communication equipment fails, staff will use phones, walkie-talkies, or signal other shelter locations for help.

5. Caring for the Animals After the Evacuation

Procedure: Once outside, animals are taken to a safe, designated area away from the building to wait for further instructions from the fire department. Staff are responsible for checking on the animals' health and safety.

Contingency Plan: If the outdoor area becomes unsafe (for example, due to smoke or further fire hazards), a backup shelter has been arranged, where animals can be temporarily relocated.

6. Fire Safety Equipment and Training

Procedure: All shelter staff are regularly trained on how to use fire extinguishers and perform basic first aid in case of injury.

Contingency Plan: If a fire extinguisher is unavailable or ineffective, staff are trained to use fire blankets or other methods to suppress small fires.

Recap Questions

1. Identify three essential elements that must be considered in emergency and contingency plans for animal management.

2. List three specific emergency situations that animal management staff should be prepared to handle, and describe one procedure or action that could be taken in each situation.

3. Explain why prioritising animal welfare is particularly important in emergency situations in animal management.

4. Explain the purpose of the Health and Safety at Work Act 1974 in the context of emergency planning within the animal management sector.

5. List three possible consequences of failing to comply with RIDDOR in the animal management sector.

6. Name two pieces of legislation, other than RIDDOR, that affect emergency procedures in animal management and briefly describe their relevance.

7. List two emergency situations in animal management that require specific procedures and contingency plans to ensure safety.

8. Explain one control measure used to minimise health and safety risks during an emergency in an animal facility.

9. Why is regular training and conducting emergency drills important in the animal management sector?

10. What are three key aspects of effective communication during an emergency situation in an animal management facility?

11. List two consequences of not having contingency and emergency plans in place in an animal management facility. Explain why these consequences could be serious for the business.

12. Describe the importance of regular staff training in animal and human first aid and fire safety in an animal care facility. Give one example of how this training directly benefits the facility.

13. Which three organisations must be notified during an emergency in the animal management sector, and what is the specific role of each?

14. Identify two potential dangers to the public that may arise during an emergency in the animal management sector and specify which service should be notified for each danger.

15. What is the most critical initial action to take when responding to an emergency in an animal care facility, and why is it important?

16. Describe two common emergency situations that might occur in wildlife or farm parks and outline the key procedures to manage them.

17. What are the main emergency concerns in aquariums, and how should staff respond to mitigate risks?

1.3 Principles of safe manual handling and their application when working in the animal management sector

Legislation and regulations

Safe manual handling is critical in the animal management sector to ensure the well-being of employees, employers, and the animals under care. Understanding relevant legislation, implementing proper techniques, and fostering a safety-first culture help prevent injuries, reduce risks, and comply with legal requirements.

Key legislation affecting manual handling

1. Health and Safety at Work Act 1974 (HASAWA)

Overview: This foundational piece of legislation sets out the duties of employers and employees to ensure workplace safety.

Employer Responsibilities:

- Provide a safe working environment, including proper training and suitable equipment.
- Conduct risk assessments to identify and mitigate hazards.

Employee Responsibilities:

- Follow provided training and safety guidelines.
- Report hazards or unsafe practices.

Relevance to Manual Handling:

- Employers must provide appropriate equipment (e.g., animal lifting aids) and ensure that staff are trained to handle animals safely, minimising risks of injury.

2. Manual Handling Operations Regulations 1992 (MHOR)

Overview: These regulations specifically address manual handling tasks, outlining the need to minimise risks associated with lifting, carrying, and moving objects or animals.

Key Provisions:

- Avoidance of Hazardous Manual Handling: Employers must avoid manual handling where possible by using mechanical aids (e.g., trolleys, harnesses).

- Risk Assessments: Employers must assess risks when manual handling is unavoidable, considering load weight, size, and environment.

Employee Duties:

- Employees must adhere to safety instructions and report issues that could lead to injuries.

Relevance to Animal Management:

- Animal handlers frequently lift or move animals and equipment. For example, moving heavy feed bags or restraining large animals must follow proper risk-assessed techniques.

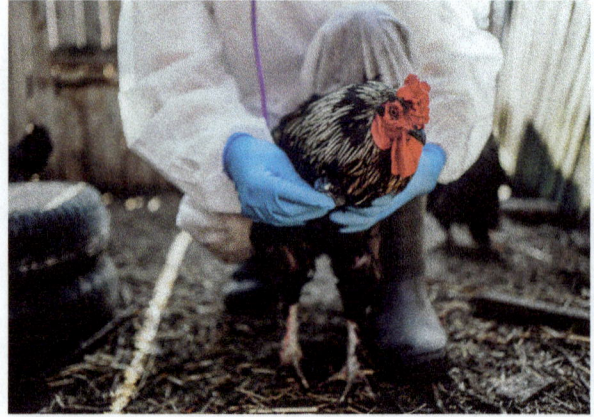

Important terms!

Manual Handling: Moving or lifting loads (including animals).

Load: The object or animal being moved.

3. Reporting of Injuries, Diseases and Dangerous Occurrences Regulations (RIDDOR)

RIDDOR ensures that specific workplace incidents are reported to the Health and Safety Executive (HSE).

Reportable Incidents Include:

- Injuries from manual handling, such as strains or sprains.

- Dangerous occurrences, such as equipment failure while moving animals.

Employer Obligations:

- Record and report incidents promptly, enabling preventive actions and compliance with legal requirements.

Relevance to Manual Handling:

- Reporting manual handling injuries highlights problem areas and informs improvements to prevent recurrence. For instance, if a handler sustains an injury while lifting a large dog, an investigation may lead to the introduction of hoisting equipment.

Why legislation and regulations matter

Promotes Safety: Reduces risks to employees and animals during handling tasks.

Legal Compliance: Avoids penalties and legal action against businesses.

Improved Workplace Practices: Encourages the adoption of safer, more efficient handling techniques.

Injury Prevention: Protects employees from common manual handling injuries such as back strains, muscle tears, and fractures.

Animal Welfare: Ensures that animals are moved or restrained safely, minimising stress and potential harm.

Roles, responsibilities and consequences

Employer responsibilities

Risk Assessments

- Identify and minimise risks associated with manual handling activities.

- Include tasks such as cleaning enclosures, lifting bedding, or moving animals.

Provide Equipment:

- Supply lifting aids like wheelbarrows or hoists to reduce manual handling strain.

Employee Training:

- Ensure employees are trained in safe manual handling techniques.

- Tailor training to the specific risks associated with working with animals, such as unpredictable movements.

Employee responsibilities

Follow Procedures:

- Adhere to training on manual handling and use the equipment provided.

Personal Awareness:

- Avoid taking unnecessary risks, such as

Consequences of non-compliance

Compliance with the law is critical. Failing to follow manual handling legislation can result in serious consequences for both employers and employees.

For employers:

- **Prosecution**: Breaches of MHOR, HSWA, or other regulations can lead to legal action.

- **Fines**: Companies found negligent may face substantial financial penalties.

- **Reputational Damage**: Non-compliance can harm a business's reputation, affecting client trust and employee morale.

- **Health and Safety Orders**: Businesses may be subject to improvement notices or prohibition orders.

> **Remember!**
> Legislation such as HASAWA, MHOR, and RIDDOR ensures safety and compliance, while proper handling techniques minimise risks. Developing these skills and knowledge will prepare you to work safely and effectively.

> **Important terms!**
> Musculoskeletal Disorders: Injuries affecting muscles, joints, or bones.
> Compliance: Following laws and regulations.

Monitor Compliance:

- Regularly review manual handling procedures and ensure employees adhere to them.

attempting to lift weights beyond personal capacity.

Report Issues:

- Inform employers about broken equipment or unsafe practices.

For employees:

- **Personal Injury**: Unsafe manual handling can cause musculoskeletal disorders, such as back injuries or sprains.

- **Loss of Employment**: Repeatedly failing to adhere to manual handling policies may lead to disciplinary action or dismissal.

- **Reduced Quality of Life**: Long-term injuries can impact both professional and personal life.

Understanding and applying **safe manual handling principles** is **critical in the animal management sector**. Both employers and employees have responsibilities to ensure health and safety, supported by comprehensive legislation. Failure to comply can lead to serious consequences, making proper training and awareness of these principles a top priority.

Principles of safe lifting techniques

Manual handling is a critical aspect of animal management, ensuring both the safety of individuals and the welfare of animals. Adhering to the Health and Safety Executive (HSE) guidelines and employing safe lifting techniques help minimise injury risks. Below are the key principles and their practical application in animal care.

Key principles of safe lifting techniques

1. Task

Before lifting any load, it is important to evaluate the nature of the task. This involves assessing the specifics of the lift to ensure it can be performed safely.

Plan the lift:
You should consider whether the item to be lifted can be divided into smaller, more manageable parts. For example, instead of attempting to carry a large, heavy feed bag (which could cause strain), it's better to split the feed into smaller bags or containers. This reduces the overall weight you're lifting at once, making the task safer.

Consider the distance and any obstacles:
Evaluate how far the load needs to be moved and the route you will take. If you need to lift and carry a heavy animal carrier across a long distance, obstacles such as steps, uneven floors, or tight doorways should be considered in advance. Plan your movement path to avoid tripping hazards, and clear the space where you will be working.

2. Load

The load itself plays a crucial role in determining whether the lift can be done safely. Proper assessment of the load is vital for ensuring you don't strain yourself.

Assess weight, size, and stability:
Before lifting, check how heavy and bulky the load is. For example, lifting a large, unstable animal carrier that could shift or tip over requires more caution than lifting a stable, evenly packed bag of food. A heavy load, such as a large bale of hay, should be assessed for balance and safety to prevent it from tipping during movement.

Ensure the load is within your capacity:
If the load is too heavy or awkward to handle, use mechanical aids, such as a trolley, hoist, or lifting straps. For example, rather than manually lifting a large dog kennel, a trolley or a lift should be used to avoid injury. Always ask for help if needed.

3. Working environment

The working environment can have a significant impact on how safe a lift is.

Check for hazards:
Ensure the area is free from potential hazards such as slippery floors, uneven surfaces, or poor lighting. If the floor is wet, there's a higher risk of slipping during the lift. In the case of animal care, ensure there are no scattered tools or equipment that could cause trips or falls while carrying animals or equipment.

Clear clutter:
A cluttered work area can lead to accidental injuries. For example, moving large feed containers or lifting animal carriers is much safer when the floor is clear of unnecessary items. If the workspace is tight, you may need to reposition equipment or organise the area before lifting.

4. Individual capability

Knowing your individual capability is crucial to prevent overexertion or injury.

Know your limits:
Only attempt to lift what is within your physical ability. For example, if you're lifting an animal or piece of equipment that feels too heavy, consider using a lifting aid or ask a colleague for assistance. For instance, carrying a large animal crate alone might not be feasible for everyone – using a two-person lift technique or hoist could be safer.

Seek help:
Never hesitate to ask for help when lifting a heavy or awkward object. If you're moving an injured or large animal, another person can help you ensure both the safety of the animal and yourself.

Proper training:
Ensure you have been trained in safe lifting techniques. Training may include proper posture, understanding how to lift with your legs rather than your back, and how to coordinate lifts with a team. Training is especially important in animal management, where safety is essential not only for humans but also for the animals involved.

5. Other factors

Maintain a neutral posture:
Always try to maintain neutral posture by keeping your back straight, shoulders back, and bending your knees when lifting. For example, when lifting a small animal crate, squat down to lift the crate rather than bending at the waist. This helps protect your back and reduces the risk of strain.

Avoid twisting or overreaching:
Never twist your body when lifting, as this can lead to back injury. Instead, turn your whole body with the load, keeping your feet planted and your movements controlled. Similarly, avoid overreaching while lifting, as this can cause loss of balance or strain.

Clothing, PPE, or machinery interference:
Ensure that any PPE (e.g. gloves, boots, protective clothing) or tools are not restricting movement during the lift. Tight clothing or heavy gloves might impair your ability to grip and carry the load. Also, check that any nearby machinery or equipment won't interfere with the lifting process – ensure there's enough space for you to move freely.

Application in animal management

Safe manual handling techniques are vital when lifting heavy items like feed bags, restraining large animals, or transporting animals in carriers. Adopting these practices minimises risks to handlers and ensures animal welfare.

• •

Other considerations

Importance of considering individual capability when manually lifting objects

Before lifting or moving objects in the animal management sector, it is important to assess the physical capabilities of the individual involved. This includes factors such as:

*   **Physical Strength and Fitness**: Ensure the person is physically capable of lifting the load. Lifting heavy objects without sufficient strength can lead to serious injuries, such as back strains or musculoskeletal disorders.

*   **Experience and Training**: People with training in safe lifting techniques are less likely to injure themselves compared to those who are untrained. Always ensure that the individual is familiar with proper lifting techniques, such as bending at the knees rather than the back.

*   **Health Conditions**: Consider whether any individual has pre-existing health issues, such as back problems, which might increase their risk of injury during manual handling.

Importance of considering the size and shape of objects

The size and shape of the objects being moved directly affect how they should be handled. For example:

*   **Size**: Large or bulky items may be difficult to carry alone and can obstruct the view, leading to trips or falls. Lifting items that are too large could also cause strain on the body if the person is not positioned correctly.

*   **Shape**: Irregularly shaped objects can be more difficult to grip and may require special handling techniques or equipment to prevent accidents. Items with sharp edges should be handled carefully to avoid cuts or punctures.

*   **Weight Distribution**: Uneven weight distribution in an object can cause imbalance, which may lead to accidents. It is important to assess how weight is distributed and whether assistance is needed.

When to ask for help

When handling large, heavy, or awkwardly shaped loads, including machinery, it is crucial to know when to ask for help. By understanding the five key factors and applying these principles, you can reduce the risk of injury while maintaining safety standards in the animal management sector.

Practical tips for animal care

- Use feed buckets with ergonomic handles.
- Employ rolling carts for heavy bedding or feed bags.
- Work with colleagues when lifting or restraining large animals.

> **Important terms!**
> Capability: An individual's physical ability to handle objects safely.
> Lifting Techniques: Correct methods to prevent injury, such as bending the knees and using the legs.
> Assistance: Seeking help when a task is too difficult or dangerous to do alone.

Real-World Example: Principles of Safe Manual Handling in the Animal Management Sector

You're a new animal care assistant at a wildlife rescue centre, where you work closely with injured and orphaned animals. One day, you need to safely transport a small but injured rabbit from its recovery pen to the veterinarian for treatment. The rabbit has been recovering from a leg injury, and while it's calm, it's still important to handle it safely to prevent further stress or injury.

Applying safe manual handling principles

Assess the Risk

Before lifting or handling the rabbit, you take a moment to assess the situation. The rabbit is small, but injured, so you know that you need to minimise any jarring movements. The environment is clean, but there may be obstacles like food bowls or bedding that could trip you, so you plan your movements carefully.

Plan

Since the rabbit is small and fragile, you plan to use a soft towel to gently pick it up. This will provide both support and protection to prevent injury to the animal. You also ensure that the transport box for the rabbit is nearby and easy to access, so you won't have to hold the rabbit for longer than necessary.

Keep a Straight Back, Bend Knees

You carefully squat down with your back straight, bending at the knees, not the waist. This technique is key in avoiding strain on your back muscles. You scoop up the rabbit gently, supporting its body from underneath with the towel, ensuring that it feels secure in your arms.

Maintain a Firm Grip

You hold the towel around the rabbit's body securely but gently, avoiding any pressure on its injured leg. By maintaining a firm, balanced grip, you can safely carry the rabbit without the risk of dropping it or causing further injury.

Communicate with Others

On your way to the vet's office, you come across a colleague. You calmly inform them that you're transporting the rabbit and ask for a hand if you need it in case of any further challenges. Good communication is vital, especially when handling live animals, as they can be unpredictable.

Minimise the Distance

You take the most direct route to the vet's office, avoiding unnecessary steps or detours. This reduces the time the rabbit has to be handled, helping to avoid stress and injury.

Recap Questions

1. What are the three key pieces of legislation relevant to safe manual handling in the animal management sector?

2. Explain how the Manual Handling Operations Regulations 1992 (MHOR) impact the responsibilities of employers and employees when handling animals or equipment.

3. Why is risk assessment considered a key principle in ensuring safe manual handling in the animal management sector? Provide two examples of its application.

4. Define a contingency plan and explain its importance in the context of animal management facilities.

5. What are two key responsibilities of an employer under the Manual Handling Operations Regulations 1992 (MHOR) in the animal management sector?

6. List three consequences an employer might face if they fail to comply with manual handling legislation.

7. Describe one safe manual handling technique and explain why it is important in the animal management sector.

8. What are the five key factors to consider according to the HSE directive when performing manual handling tasks?

9. Why is it important to assess the working environment before lifting a load?

10. What should you do if a load is too heavy or awkward to lift on your own?

11. Why is it important to consider an individual's physical capability before attempting to lift or move objects in the animal management sector?

12. When is it necessary to summon assistance to move objects or loads? Name and explain two of the five factors that should be assessed when deciding whether help is needed.

13. How do the size and shape of objects impact manual handling and the need for assistance?

Practice Questions

1. Explain the importance of maintaining a clean and organised workspace in animal care. (4 marks)

2. Describe two key steps involved in conducting a risk assessment in an animal management setting. (4 marks)

3. Identify and explain two types of hazards commonly found in animal management facilities and the potential risks associated with each. (6 marks)

4. List three responsibilities of an employer in maintaining health and safety in an animal care environment. (6 marks)

5. Explain why it is essential to follow proper lifting techniques when handling large animals or heavy equipment in an animal care facility. (3 marks)

6. List and explain two legal responsibilities that animal care staff have in ensuring health and safety in the workplace. (4 marks)

7. Describe two environmental hazards that may arise in an outdoor animal enclosure and how to mitigate these risks. (5 marks)

8. Identify three potential risks when working with unfamiliar animals and provide one example of a control measure for each. (6 marks)

2.1 Waste management principles in the animal management sector

Legislation and regulations

The five Rs of waste management

Proper waste management is an essential aspect of animal care environments, ensuring that waste is disposed of efficiently, sustainably, and safely. The five Rs – refuse, reduce, reuse, repurpose, and recycle – are principles that help manage waste effectively. These principles are directly applicable when dealing with waste in animal care facilities.

1 Refuse

Refusing waste involves avoiding products or materials that contribute to unnecessary waste production. In the animal management sector, this could involve refusing single-use plastics or non-biodegradable materials that are difficult to recycle. For example, opting for reusable containers and avoiding over-packaging in animal care products.

2 Reduce

Reducing waste refers to minimising the amount of waste generated in the first place. In animal management, this can be achieved by ordering supplies in bulk to reduce packaging or by using products that have a longer lifespan. Reducing waste also means properly managing resources like food, ensuring that animals are fed the appropriate amounts to prevent food waste.

3 Reuse

Reusing materials involves finding new uses for items that would otherwise be discarded. For example, reusing animal bedding or old blankets as cleaning rags or donating unwanted equipment to local animal shelters. This helps cut down on the demand for new resources and reduces the burden on landfill sites.

4 Repurpose

Repurposing means transforming waste items into new products. For example, using leftover food or organic waste from animal care as compost or mulch for gardening. This minimises waste and turns it into a resource, benefiting the environment.

5 Recycle

Recycling involves converting waste into reusable materials, such as plastics, paper, and glass. In the animal management sector, recycling could involve separating waste into different categories for appropriate disposal or using recycling bins in facilities to ensure waste is processed correctly.

> **Important terms!**
>
> **Waste management**: the process of handling, storing, and disposing of waste properly.
>
> **Five Rs**: a set of principles to manage waste: refuse, reduce, reuse, repurpose, and recycle.
>
> **Legislation**: laws or rules created by authorities to govern activities, like waste disposal.
>
> **Controlled waste**: waste that is regulated because it could be harmful to health or the environment.
>
> **Animal by-products**: parts of animals, like carcasses or body parts, that aren't used for food but must be safely disposed of.
>
> **Hazardous waste**: waste that can be dangerous to people or the environment, like chemicals or contaminated materials.
>
> **Environmental protection**: efforts to protect the environment from harm caused by human activities, including pollution and waste.
>
> **Nitrate vulnerable zones (NVZ)**: areas where the use of animal manure and fertilisers is limited to protect water quality from pollution.

Reduce   Reuse   Recycle

Key requirements of legislation and regulation affecting waste management

In the animal management sector, understanding and adhering to various regulations is essential to ensure proper waste management and compliance with environmental standards. Below are key pieces of legislation that affect waste management in the UK:

Controlled Waste (England and Wales) Regulations 2012

This regulation **classifies waste as either household, industrial, or commercial**. It outlines how different types of controlled waste should be handled, stored, and disposed of. In the animal management sector, this regulation is crucial when dealing with waste from animal care operations, such as medical waste, animal bedding, and food waste.

Nitrate Vulnerable Zones (NVZ) EC Directive

This EU directive aims to **reduce water pollution caused by nitrates** from agricultural sources. Animal waste, including manure, is a significant source of nitrates. Nitrate vulnerable zones (NVZ) regulations set limits on the amount of nitrogen fertiliser and manure that can be spread on land in these zones. Understanding these regulations helps animal management facilities comply with environmental protection guidelines.

Environmental Protection Act 1990

This Act provides the **framework for the management and disposal of waste in the UK**. It includes provisions for the control of hazardous waste, waste management licensing, and the general duty of care when handling and disposing of waste. Animal management facilities must ensure that their waste disposal methods comply with these requirements to prevent environmental harm.

Environment Agency rules and associated derogations

The **environment agency** in the UK is responsible for overseeing waste management and environmental protection. It issues rules and guidelines related to waste handling, including how animal waste should be treated. Derogations refer to specific exceptions granted under certain conditions, allowing waste to be handled differently based on the type or purpose of the waste.

The Animal By-Products (Enforcement) (England) Regulations 2013

These regulations govern the disposal and processing of animal by-products (ABPS). ABPS must be handled and disposed of according to specific rules to prevent disease transmission. For example, pet carcasses, body parts, or animal waste from a veterinary practice must be disposed of using approved methods, like rendering or incineration.

The Waste (England and Wales) Regulations 2011

These regulations set out the **responsibilities of individuals and businesses in England and wales for managing waste**, including the requirement to segregate waste streams. In animal care, this is particularly relevant for ensuring that hazardous animal-related waste is separated from general waste to prevent contamination.

Clean Air Act 1993

The clean air act **regulates air pollution, particularly the burning of waste materials**. In the context of animal care, this can apply to waste disposal methods like incineration. Animal management facilities must comply with emission standards to avoid air pollution, especially when burning animal waste or by-products.

> ### Remember!
> Understanding the waste hierarchy outlined in the Waste (England and Wales) Regulations 2011 is key. These regulations set a clear framework for managing waste in a way that minimises environmental impact.

Specific waste types

Effective waste management is a critical part of maintaining high standards of hygiene, health, and safety in the animal management sector. Whether it involves managing waste produced by animals, the facilities housing them, or the medical treatments provided, all forms of waste must be categorised, handled, and disposed of responsibly.

Classification of waste

In animal management, waste is categorised into various types based on its **composition, potential impact,** and **required handling procedures.** These categories help determine how the waste should be disposed of and whether it requires specific treatment or containment methods. The main types of waste are as follows:

- **Consumable waste**: items that are used and discarded in the normal course of business operations, such as food packaging, animal bedding, and food scraps. These need to be carefully disposed of to prevent contamination or attraction of pests.

- **Non-consumable waste**: items that are not used up but instead discarded, like broken equipment, tools, or containers that are no longer fit for purpose.

- **Commercial waste**: commercial waste includes waste produced by business activities such as packaging materials, discarded tools, and other commercial items that require proper disposal or recycling.

- **Domestic waste**: domestic waste is generated in a domestic setting, often including household waste, such as paper, plastics, and organic matter from food preparation. This type of waste must be segregated and disposed of to prevent cross-contamination with other types of waste.

- **Laboratory waste**: this includes waste produced in veterinary or animal research settings, such as test tubes, petri dishes, and other laboratory equipment that may be contaminated or hazardous.

- **Pharmaceutical waste**: this includes unused, expired, or discarded pharmaceutical products, such as medications, syringes, and other medical supplies. Improper disposal of pharmaceutical waste can harm the environment and public health.

- **Infectious clinical waste**: waste that contains or is contaminated with pathogens, such as infected bandages, gloves, and needles. This type of waste requires special treatment to prevent disease transmission.

> **Important terms!**
>
> Commercial waste: waste produced by businesses, such as shops, offices, and factories.
>
> Domestic waste: waste generated from households, including everyday items like food scraps, packaging, and old items.
>
> Hazardous waste: waste that can be harmful to people, animals, or the environment, such as chemicals or toxic materials.
>
> Biohazardous waste: waste that contains harmful biological materials, like medical waste, animal waste, or anything that could spread disease.
>
> Licensed carriers: companies or individuals that are legally authorised to transport waste, including hazardous or controlled waste.
>
> Controlled waste: waste that is regulated by law due to its potential danger to health or the environment, like hazardous or biohazardous waste.

- **Non-infectious clinical waste**: waste that has been used in a clinical setting but does not contain pathogens. Examples include bandages, gloves, and disposable instruments that are not contaminated with infectious agents.

- **Organic waste**: this refers to biodegradable waste such as animal manure, uneaten food, and plant-based materials. It must be managed in a way that prevents pollution and maintains sanitation.

- **Inorganic waste**: non-biodegradable waste, such as plastics, glass, and metals, that can often be recycled or disposed of in landfills. These materials can have long-lasting environmental impacts if not disposed of properly.

- **Recyclable waste**: materials that can be processed and reused, such as paper, plastic, and certain metals. These should be separated from other types of waste to minimise their environmental impact.

- **Non-recyclable waste**: waste that cannot be reused or recycled, such as certain types of plastics or contaminated materials, which require disposal in landfills or through incineration.

- **Hazardous waste**: Hazardous waste is dangerous or potentially harmful to humans, animals, or the environment. This includes chemical waste, sharp objects like needles, and waste with toxic properties.

- **Biohazardous waste**: biohazardous waste poses a risk to living organisms due to contamination by biological agents, such as blood, tissues, and waste from infected animals.

Types of waste requiring specific actions

Certain types of waste produced in animal management environments require particular handling to ensure they do not pose a risk to public health or the environment. These include:

- **Controlled waste.** Controlled waste refers to waste that is regulated by law due to its potential environmental or health hazards. This includes hazardous materials, clinical waste, and waste that requires special disposal arrangements. Licensed carriers are responsible for removing and disposing of controlled waste in accordance with environmental regulations.

- **Hazardous waste.** Hazardous waste includes substances that can cause harm to humans, animals, or the environment, such as chemicals, medicines, and used syringes. It must be segregated, stored in secure containers, and transported by licensed carriers to specialised disposal facilities. The disposal of hazardous waste is governed by strict regulations to prevent harm.

- **Animal waste.** This category includes waste produced by animals, such as hair, feathers, fur, and faeces. Animal waste must be contained in a way that prevents contamination and the spread of disease. In some cases, it can be composted or incinerated to minimise its environmental impact.

- **Clinical waste.** Clinical waste includes items like needles, used dressings, and other medical supplies that are contaminated with bodily fluids or pathogens. Clinical waste must be stored in colour-coded, leak-proof containers, and disposed of in designated facilities. Infectious clinical waste requires more stringent handling than non-infectious clinical waste.

- **Offensive waste.** Offensive waste includes items such as faeces, urine-soaked bedding, and waste that is unpleasant but not necessarily infectious. Although it is not a health hazard, offensive waste must still be managed properly to maintain hygiene and prevent unpleasant odours or pest infestations.

- **Biohazardous waste.** Biohazardous waste includes materials contaminated with biological agents, such as blood, tissues, or fluids from infected animals. This type of waste requires careful handling to avoid exposure to harmful pathogens. It must be disposed of in biohazard bags and treated in accordance with biohazardous waste protocols.

> ### Remember!
> The proper disposal of hazardous and biohazardous waste is particularly important, as improper handling can result in the spread of infectious diseases, environmental contamination, or harm to humans and animals. Licensed carriers ensure that all waste is transported to the appropriate facilities for safe disposal, in compliance with legal regulations.

Purpose and consequences

Waste management is a critical component of animal care and management. It ensures a clean, safe, and healthy environment for animals, staff, and visitors. Effective waste management is essential in animal management settings, including animal shelters, veterinary practices, farms, and zoos.

Purpose and benefits of waste management plans

Waste management plans are essential to ensure that waste is handled in a systematic, effective, and safe manner. These plans help organisations address and manage waste generation in a way that minimises risks to health, safety, and the environment.

The purpose of waste management in animal environments includes:

- **Health and safety**: proper waste disposal prevents contamination, reduces the spread

> ### Important terms!
> Zoonotic diseases: diseases that can be transmitted from animals to humans, often through improper waste handling.
>
> Environmental Protection Act 1990: a UK law that sets out responsibilities for waste management and pollution control.
>
> Manure: animal waste, especially from livestock, that can be hazardous to the environment if not disposed of properly.

of diseases, and creates a safer environment for both animals and humans. This is especially important in veterinary practices and animal shelters where animals may have diseases that can be transmitted through waste.

- **Environmental protection**: waste, especially hazardous waste like animal waste, medicines, and chemicals, can have detrimental effects on the environment if not handled properly. By managing waste responsibly, facilities help protect ecosystems and wildlife.

- **Compliance with legislation**: many countries have regulations that govern waste disposal, including specific guidelines for hazardous waste and biological waste. Waste management plans ensure that organisations comply with these rules.

The benefits of implementing effective waste management plans in animal environments are:

- **Reduction of contamination risks**: a proper waste management system ensures that harmful substances do not spread into the environment or animal habitats.

- **Improved animal welfare**: clean and hygienic environments support the overall health of animals, reducing the risk of disease transmission and promoting good welfare.

- **Cost savings**: an efficient waste management plan helps reduce costs by optimising waste disposal and recycling processes. It can also reduce the need for costly cleaning and sanitisation services due to contamination.

Financial implications of ineffective waste management

Ineffective waste management can result in significant financial costs for animal care organisations. These costs arise from:

Fines and penalties: failure to comply with waste disposal regulations can lead to substantial fines. For example, improper disposal of hazardous waste can result in hefty financial penalties.

Increased waste disposal costs: poor waste segregation or improper waste handling can lead to more expensive disposal methods. For instance, failing to separate recyclable materials from general waste could mean higher disposal fees.

Cost of environmental damage: when waste is not properly managed, it may

result in environmental damage, such as soil contamination or water pollution, leading to expensive cleanup efforts.

Insurance costs: if an animal management facility is found negligent in waste handling, its insurance premiums could increase, further impacting the organisation's financial stability.

> ### Remember!
> Ineffective waste management not only risks health and safety but also has direct financial consequences. It is critical for organisations to implement effective strategies to avoid such costs.

Legal and regulatory requirements in waste management

Some of the key legal requirements include:

- **Environmental protection laws**: these laws regulate how hazardous and non-hazardous waste should be treated. For instance, the Environmental Protection Act 1990 in the UK outlines the responsibilities for waste management and disposal. It includes rules for handling animal waste, pharmaceutical waste, and chemicals.

- **Animal waste management regulations**: specific rules govern the disposal of animal waste, such as manure, bedding materials, and animal carcasses. These guidelines aim to reduce the spread of zoonotic diseases and prevent contamination.

- **Health and safety regulations**: legislation such as the Health and Safety at Work Act 1974 requires that employers protect the health and safety of employees, including the safe disposal of waste.

Failing to comply with these regulations can have severe consequences, including:

- **Prosecution**: non-compliance can lead to legal action, resulting in prosecution or civil lawsuits. This may involve fines, closure of the facility, or imprisonment for responsible parties.

- **Reputation damage**: non-compliance or ineffective waste management can seriously damage an organisation's public reputation. The public may lose trust in an organisation that is seen as careless with waste disposal, especially when it concerns animal health.

- **Environmental harm**: improper disposal of waste can lead to contamination of soil, water, and air, damaging local ecosystems and harming wildlife. This can result in costly fines, legal action, and long-term environmental damage.

- **Legal action**: organisations may face legal prosecution for improper waste disposal. The severity of the legal consequences will depend on the nature of the waste and the level of negligence involved.

Real-world example: Waste management in a veterinary clinic

Imagine you're working at a veterinary clinic that handles the care of animals, from routine check-ups to surgery. The clinic generates a variety of waste, including animal waste (like faeces and urine), medical waste (such as used bandages, syringes, and medications), and general waste (like plastic packaging and paper towels). Proper waste management is essential to ensure a clean and safe environment for both the animals and staff.

Here's how the principles of waste management work in this scenario:

Segregation of waste

At the veterinary clinic, all waste is carefully separated into different categories:

- Medical waste (sharps, used surgical instruments) goes into biohazard containers.
- Animal waste (faeces, bedding) is placed in designated compostable bags or bins.
- General waste (packaging, plastic) is sorted for recycling or disposal in regular bins.

Minimising waste generation

The clinic implements practices to reduce waste. For example, reusable surgical instruments are sanitised and reused, reducing the need for disposable ones.

They also use eco-friendly cleaning products that come in refillable containers to minimise plastic waste.

Disposal and recycling

Medical waste is disposed of through a licensed medical waste disposal service, ensuring that it is handled safely and in compliance with regulations.

- Recyclable items, like paper and plastic packaging, are sent to a recycling facility.
- Animal waste that cannot be composted is properly disposed of in a way that minimises environmental impact.
- Education and training: the staff are regularly trained on how to properly handle and dispose of different types of waste, ensuring that everyone is aware of their role in waste management.
- New employees are educated on the clinic's waste management policies to ensure safe

Recap Questions

1. What are the five Rs of waste management, and how can they be applied in the animal management sector?

2. Explain the main purpose of the environmental protection act 1990 in relation to waste management in the animal care sector.

3. What is the role of the animal by-products (enforcement) (England) regulations 2013, and why is it important in the handling of animal waste?

4. What is the main difference between clinical waste and hazardous waste in the animal management sector?

5. List three types of waste that require specific actions and describe one action required for each.

6. Why is it important for animal management businesses to use licensed carriers for waste disposal?

7. What are two key benefits of implementing an effective waste management plan in animal environments?

8. List three potential consequences of failing to comply with waste management regulations in animal management settings.

9. What is the purpose of the environmental protection act 1990 in relation to waste management in animal care facilities?

Practice Questions

1. Define the five Rs of waste management and explain the significance of each in reducing the environmental impact of waste in the animal management sector. (5 marks)

2. Describe the process and specific actions required to manage hazardous waste in an animal care environment. Include why these actions are necessary. (5 marks)

3. Explain the legal requirements and regulations that impact the disposal of clinical waste in the animal management sector. (5 marks)

4. What are the potential financial and environmental consequences of poor waste management in an animal care facility? (5 marks)

5. Describe the difference between controlled waste and non-controlled waste, and explain how each type must be handled according to regulations. (5 marks)

3.1 Biosecurity measures

Biosecurity is the practice of preventing the introduction and spread of infectious diseases, pests and invasive species. It is vital for animal welfare, human health and safety and the environment. Biosecurity takes place at multiple levels, from small collections and individual farms to protecting national borders.

In this topic you will learn different methods of biosecurity and why they are essential in the animal industry.

Purposes of biosecurity in the animal management sector

There are several different aims of biosecurity within an animal collection:

Prevent / control the spread of disease

The most important aim of biosecurity is to **contain disease outbreaks** and prevent the disease from spreading, both inside the collection and outside of it. Animals do become unwell, and when it happens, it is the responsibility of everyone working within the organisation to ensure that animals and people are not put at risk.

Prevent / control the introduction of new diseases

Although it may not be possible to prevent all diseases occurring within an animal collection, **preventing the introduction of new diseases** can be achieved through vigorous application of biosecurity control measures. The aim here is to ensure that new animals that are brought into the collection are free from disease before they are allowed access to animals already housed there.

Maintain animal welfare

Ensuring that **animal welfare is of the highest standard** will reduce the risk of animals developing illnesses. Providing appropriate housing, enrichment and diet supports overall animal health. In addition to this, implementing measures to reduce stress (which can compromise the immune system) will decrease disease susceptibility. Regular veterinary checks, along with prophylactic care, can reduce the risk of disease even further.

Maintain human health

Certain diseases can pass from animals to humans. These zoonotic diseases can pose a risk to human health - therefore **protocols must be put in place to reduce this risk.** If doctors discover that a person has been infected with a zoonotic disease, they will often inform the local authorities so that further measures can be put in place to protect people. Diseases that must be reported to the local authorities are known as notifiable diseases.

Maintain species

There are many benefits to **maintaining species in their natural habitat.** Each ecosystem is a complex mix of species all working together. Increasing new species or reducing the numbers of endemic species can be disastrous for the ecosystem.

Conservation programmes

Implementing biosecurity measures in wildlife reserves and breeding programs is essential to **protect endangered species from disease.** Animals that are to be released from captive breeding programmes must be free from disease to ensure they are not acting as vectors for pathogens. Introducing disease into the natural environment could be disastrous for wild populations.

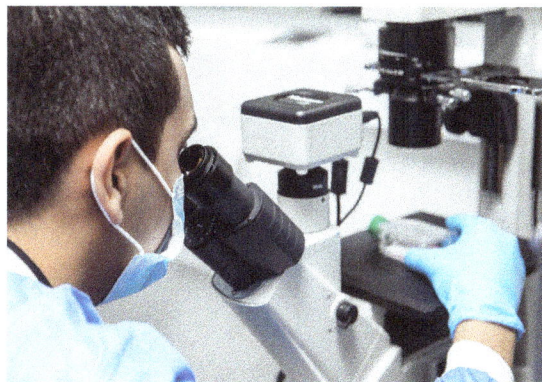

Figure 3.1.1 Scientist reviewing samples

Habitat management

Maintaining healthy environments to support wildlife populations (both plant and animal) will help to prevent disease outbreaks.

Reducing invasive species and pests will prevent disease outbreaks and reduce their spread. Invasive species can include plants, animals, insects and fungi. When these species are introduced to a new area and are able to establish, they often do not have any natural predators to control their numbers, so their population grows quickly. This increases the competition for food and habitat, as well as introducing new predators that threaten endemic species.

Biosecurity monitoring

Regularly monitoring of wildlife health will help to detect and manage emerging threats to species. Capturing, testing and tagging individuals from a population can highlight any problems before they become unmanageable.

Public awareness

Educating the public on the importance of biosecurity in protecting both wildlife and captive animals can help prevent the spread of disease, particularly in areas visited by walkers and hikers. Promoting practices that minimise human impact on natural habitats can help to reduce the spread of disease. This can include things such as simple signage in key areas advising visitors to keep dogs on leads to protect vulnerable animals or ground-nesting birds or educational television series highlighting the importance of the natural world.

Asian hornet (*Vespa velutina*) is an invasive species in the UK which is putting honey bee populations at risk

Biologist setting camera traps to monitor wildlife populations

---

Important terms!

Animal collection: A group of animals that are kept for display, education, breeding, conservation or as pets.

Biosecurity: The standards and practices in place to prevent the introduction / spread of disease, pests and invasive species.

Notifiable Disease: A serious disease that must be reported to the local authority. These diseases must be reported if they are discovered in humans or animals so that measures can be put in place to prevent their spread. Examples include foot and mouth and rabies.

Endemic: This refers to any disease or species that is naturally found within a geographical area. For example, emus are endemic to Australia.

Vector: An organism that carries an infectious disease, such as an animal, a person or mosquitos.

Pathogen: A disease-causing agent such as bacteria, virus, fungus or prion.

Prophylactic: Preventative treatments such as vaccinations or worming that are used to prevent or reduce the severity of disease.

Codes of practice under Defra and APHA

The **Department for the Environment, Food and Rural Affairs** (**Defra**) is a UK government department responsible for policies and regulations related to the environment, food production, rural communities, and animal health and welfare.

Defra oversees animal welfare legislation, regulates waste management and manages national emergencies related to food, farming and the environment.

The **Animal and Plant Health Agency** (**APHA**) is an executive agency of Defra that focuses on safeguarding animals and plant health, animal welfare and public health in relation to zoonotic diseases.

Standards and organisational policies

Defra sets national standards and policies for animal health, welfare and environmental protection. These provide frameworks for managing disease outbreaks and biosecurity.

APHA enforces these standards by conducting inspections of animal collections and providing support where needed. They implement codes of practice for the welfare of farm animals, transport, disease control and biosecurity. APHA helps to maintain high standards in managing animal and plant health risks through disease surveillance and border checks.

The Defra Codes of Practice for the Welfare of Animals provide detailed advice for caring for all aspects of farm animal and pet welfare, ensuring their basic needs are met as per the Animal Welfare Act 2006. (See later in this section for more about this Act).

Implications of not following codes of practice

There are several, severe implications of not following these codes of practice.

Legal consequences

Non-compliance with codes of practice means a collection is breaking the law. This can result in fines, prosecutions or loss of business licences. If this non-compliance has led to breaches in animal welfare, prosecutions can happen under the Animal Welfare Act 2006.

Public health risks

Ignoring biosecurity measures (whether intentionally or accidentally) can lead to the spread of zoonotic diseases which can have serious impacts on public health.

Economic Impact

Non-compliance in disease management can result in trade bans, financial losses, or widespread culling of animals. In many cases of notifiable disease, if just one animal in a herd or flock is found to be infected, all animals in that herd or flock will be culled, leading to significant financial loss for farmers.

Environmental breaches (e.g., pollution) can lead to costly clean-up operations and reputational damage.

Environmental consequences

Non-adherence to environmental standards (e.g., waste disposal or pesticide use) can harm ecosystems, reduce biodiversity, and lead to legal action.

Industrial waste water released into a river

> **Important terms!**
>
> Codes of practice: Guidelines or standards that provide practice advice on how to comply with laws, regulations or ethical standards. These link closely to legislation and are in place to help animal collections meet the minimum standards required.
>
> Defra: Department for the Environment, Food and Rural Affairs
>
> APHA: Animal and Plant Health Agency
>
> Culling: Humane slaughter of animals on a large scale, often to help prevent the spread of disease

Factors influencing biosecurity

- **National/international trade**. The movement of animals, plants and goods between countries increases the risk of introducing non-native diseases, pests or pathogens. Proper biosecurity measures (e.g. health certifications and inspections) are essential to reduce risks associated with trade.

- **New technologies**. Advancements in diagnostics, surveillance and tracking systems enhance early detection and management of biosecurity risks. Gene editing and biotechnology may create risks (e.g. modified organisms) requiring strict monitoring and regulation.

- **Disease outbreak**. An outbreak of disease highlights weaknesses in biosecurity systems and prompts stricter controls. Outbreaks increase the need for containment, such as restriction zones and mass testing.

- **Disease control.** Effective disease control measures (e.g. vaccination, culling, disinfection, staff training) directly impact biosecurity by limiting the spread of infectious agents. Inadequate control can lead to wider outbreaks and compromise biosecurity.

- **Outbreak management implications**. Poorly managed outbreaks put a strain on resources, disrupt trade and damage public trust in biosecurity. Effective management (for example rapid response teams and public communication) ensures containment and reduces economic impact.

- **Border control**. Screening and inspections at borders help prevent the introduction of diseases and invasive species through

Tortoises being smuggled in an old suitcase

imports and travellers. Strict rules on import and export mean that government organisations know what species are being transported and the relevant checks can be carried out.

- **Quarantine**. Isolation of plants, animals or goods helps identify and contain potential risks before they spread. Compliance and enforcement of quarantine rules are critical to biosecurity success.

- **Illegal pet trade and smuggling**. The illegal trade in exotic species for the pet trade introduces animals that may carry zoonotic diseases or invasive pets, undermining biosecurity measures. Animal welfare is put at risk because of the capture and transport methods that people use when smuggling animals. People bringing animals into the country illegally are very unlikely to carry out health screening or quarantine procedures, which means biosecurity is at risk.

The illegal pet trade is difficult to monitor and control and therefore poses a significant risk to national biosecurity.

Zoonotic and notifiable diseases

Definition of zoonotic, anthroponotic and notifiable disease

Zoonotic is the term given to diseases that can pass from animals to humans. Examples include rabies, foot and mouth, salmonella, Lyme disease and toxoplasmosis.

Notifiable diseases are zoonotic diseases deemed a serious risk to human health. When a notifiable disease is suspected or confirmed, vets, farmers, animal keepers and doctors are required to report it to the relevant authorities so that steps can be taken to control the outbreak and prevent it from spreading. Examples include foot and mouth disease, avian influenza, bovine TB, rabies and Newcastle disease.

Anthroponotic diseases are diseases that can pass between humans or from humans to animals. These can pose a risk to animal health. Examples include tuberculosis, influenza, the common cold, and ringworm. Although these diseases may not be communicable with all animal species, some will be more at risk than others.

Role of APHA and Defra

As previously discussed Defra is a UK government department responsible for policies and regulations related to the environment, food production, rural communities, and animal health and welfare. Defra oversees animal welfare legislation, and manages national emergencies related to food, farming and the environment.

APHA is an executive agency of Defra that focuses on safeguarding animals and plant health, animal welfare and public health in relation to zoonotic diseases.

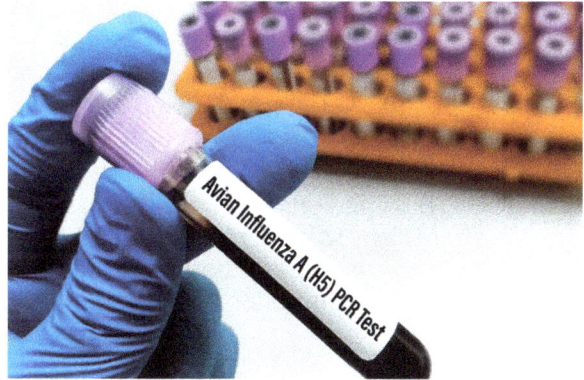

PCR tests for avian influenza, a zoonotic disease

Movement and restriction zones for notifiable disease

When a notifiable disease is suspected or confirmed, APHA will put **movement restrictions** on infected animals and people. These restriction zones can be classified as 'immediate' or 'surveillance' zones. The difference between them is their purpose, size, and the measures implemented to control and monitor disease outbreaks.

Immediate

Immediate zones are established around a suspected or confirmed outbreak of a notifiable disease. The aim is to contain the disease and prevent it from spreading. These tend to be smaller and focused on high-risk areas around the outbreak site.

Restrictions in immediate zones limit the movement of animals, the people who can access the area and vehicles entering or leaving the premises. Movement can be limited or completely restricted depending on the outbreak.

Animals within immediate restriction zones will be tested and if any are found to be carrying the disease, all animals may be culled to prevent the disease from entering the food chain.

Surveillance zones

Surveillance zones are less severe than immediate restriction zones and are used to monitor for signs of disease in a wider area. In surveillance zones, the movement of animals will be monitored but not completed restricted. Testing of animals at risk will be carried out to detect any new cases. Increased biosecurity measures may be implemented.

Biosecurity control measures

Staff and visitors' limitation

Putting **limits on staff and visitors in specific areas** can limit the spread of disease as well as allowing animals to recover in a quiet environment, uninterrupted by people who may unintentionally cause stress. Many animal collections will limit the number of staff authorised to enter the isolation and quarantine areas to limit the risk of spreading infection.

In collections where biosecurity is very high, for example laboratories, staff may be limited to one area only to ensure disease is not being transmitted between sections.

Buildings, equipment and vehicles' cleanliness/usage protocols

In biosecurity, when we think about movement, we are not just talking about the movement of animals. **Restricting the movement of equipment and vehicles** between different

areas, particularly during an outbreak, will limit the likelihood of transmission.

A lot of collections try to pre-empt this by having different equipment for different tasks. A good example would be using different cleaning equipment for different areas or having specific pieces of feeding equipment for each species or enclosure.

Animal management

Animal facilities should be designed with **controlled entry points** so that new animals do not come into contact with any others in the collection until they have passed through quarantine. Air filtration systems, and dedicated areas with additional controls such as PPE and foot dips, will help prevent new diseases from entering the population.

In addition to this, animal care **workers should be trained to recognise the first signs that an animal may be unwell**, so that it can be removed from the collection and treated as necessary. Many prey species will not show physical symptoms of illness until they are too ill to hide it. The first signs that these animals are unwell are likely to be behavioural, so it is vital that animal care workers understand how to spot these subtle indications that something may be wrong.

Control of parasites, pests and vermin

Parasites are species that live on a host. They can be:

- internal (**endoparasites** such as worms or flukes)

- external (**ectoparasites** such as fleas and ticks).

Parasites can act as vectors, passing on disease to the host animal, so it is vital they are controlled. Using schedules for prophylactic treatment can prevent infestations and reduce the likelihood or severity of ill health.

Pests and vermin are insects or animals that cause problems for animal collections.

- **Vermin** such as rats and mice can chew through enclosures and food containers, bringing disease with them. Food should be kept in secure containers that cannot be chewed and enclosures checked regularly and repaired as necessary. The use of traps and poison may be required in areas where rat and mice numbers are high.

- **Pests** such as flies can be a problem in areas where food is prepared. Ensuring high levels of hygiene, correct food storage and waste disposal will ensure numbers are kept low. It may not be possible to remove flies completely, especially in warmer weather, so the use of traps can help reduce their numbers.

It is also important to **check animals regularly**, particularly in warmer weather. Flies will lay their eggs on animals with dirty fur or a dirty fleece. These eggs hatch into maggots which then eat into the animal's skin, causing a condition called

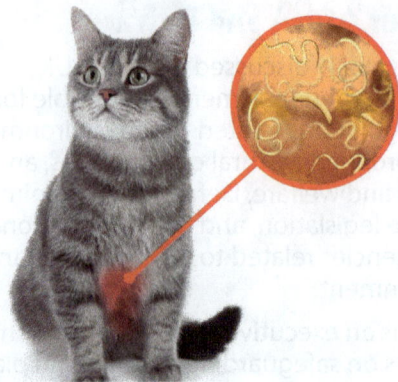

Cat with endoparasites

flystrike. High standards of hygiene and applying preventative sprays to animals will help prevent infestation.

Inspection and monitoring

Inspecting animals, enclosures and facilities regularly is vital to ensuring biosecurity. Inspections should be carried out by trained personnel at the collection according to the schedule produced. Certain organisations, such as those with a zoo licence, will be inspected by a representative of the local authority and two approved specialist inspectors. These inspections must be carried out every three years.

Recording and monitoring biosecurity is the responsibility of everyone who works in an animal collection. Keeping records of concerns, animal health, animals entering and exiting the collection, as well as birth and deaths, can help reveal patterns and identify problems before they escalate.

Regulation

In order to ensure all collections are following the same guidelines, biosecurity is covered in several pieces of legislation in the UK. This regulation ensures that minimum standards and requirements are being met.

See later in this section for details about some specific legislation

Passports and border controls (including illegal trade)

Certain species of animal have very strict controls in place for movement. An excellent example of this is cattle and horses, which require a passport which must be updated as they move. This ensures that every individual from these species is accounted for in case of a disease outbreak, and owners can be contacted swiftly when necessary.

When collections bring new species of plants or animals in, there is a chance that they could be bringing disease with them. This is especially

Rats infesting poultry nests

true if they have not been obtained from a reputable source (for example animals being illegally imported).

All new species entering a collection should undergo a full health check, including blood and faecal sampling, and remain in quarantine until they have been signed off as healthy.

Although biosecurity is important when moving animals between collections, it is vital when moving animals between countries. Each country has organisations and government departments that are responsible for border control. Their aim is to reduce the likelihood of introducing invasive species, reducing the risk of disease as well as to prevent animals being illegally imported.

Quarantine and isolation

Any animals that are unwell should be separated from the rest of the group. This will prevent the disease from spreading to others. Many collections have a dedicated **isolation and quarantine** area with increased biosecurity measures to increase the efficacy of these measures.

All animals coming into a collection should undergo a thorough health check, including faecal egg counts and blood sampling, to rule out diseases they could be carrying. If the tests come back positive, the animal can then be treated accordingly and retested at the end of the course of treatment.

Once these tests have come back negative, the animal should be monitored for a period of time to ensure that all traces of disease have cleared their system.

Cleaning protocols

Regular **cleaning and disinfection** of animal enclosures, surrounding areas, labs, equipment, vehicles and working areas will help to eliminate pathogens. Staff should be trained on the proper usage of cleaning chemicals, as dilution rates are linked to efficacy.

- Disinfectant that is too weak may not destroy all pathogens present and could lead to resistance developing
- Whereas disinfectant that is too strong may pose a risk to animal and human health.

As well ensuring enclosures and equipment are kept clean, staff should be trained on how to correctly wash their hands in order to maintain high standards of personal hygiene.

Food storage and water cleanliness

Food should **always be stored so that pests and vermin cannot gain access.** Metal feed containers will prevent mice and rats accessing and contaminating feed.

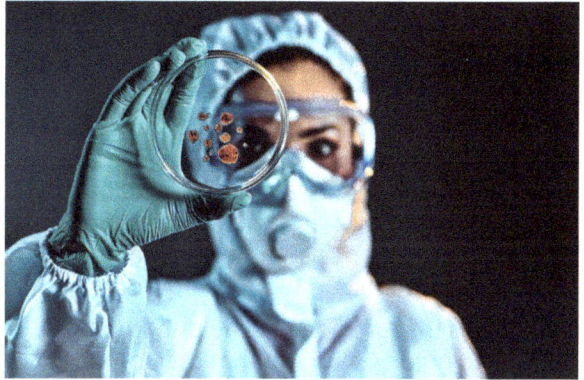

Scientist reviewing the biosecurity screening results

Keeping food refrigerated or frozen will prolong its life and maintain quality for longer. It is very important to ensure that frozen food is fully defrosted before feeding to animals.

Food bowls, puzzle feeders and any other methods used to provide food to animals should be cleaned as part of the daily feeding routine to prevent the build up of bacteria or mould.

Water sources can be natural, such as a stream running through a large enclosure or provided in a container, such as a bowl or water bottle. Natural sources of water should be checked regularly for contaminants and pollutants that have entered the water further upstream. Water bowls and bottles should be washed regularly to prevent the build up of limescale, algae or bacteria.

PPE

Personal protective equipment (**PPE**) is vital for preventing injury to humans but it also plays a role in biosecurity. Having clothing that is only worn in certain areas will prevent the spread of disease. Regular washing and disinfecting of PPE will reduce the risk of pathogens moving between areas.

When restriction zones are in place, disposable PPE is often worn in high-risk areas so that it can be removed and destroyed before leaving the area.

Exclusion zones

Exclusion zones are the most extreme version of a restriction zone. These are generally implemented in only the most severe cases of disease outbreak. Exclusion zones have strict limitations or complete prohibition on access, whereas restriction zones have regulated access to manage risks.

The purpose of an exclusion zone is to complete isolate the source of danger.

Physical barriers

Physical barriers are used in animal collections to control the movement of both animals and people. Animal enclosures are often designed with solid walls separating them, so animals in adjoining enclosures cannot make contact with each other. Likewise, solid barriers prevent visitors to animal enclosures from coming into contact with animals.

Collections that are open to the public use barriers to guide visitors so that they cannot enter restricted areas, such as isolation and quarantine or food preparation areas. This not only reduces the risk of injury but increases biosecurity as well.

Barrier nursing

Barrier nursing is the term used to describe methods that create a physical barrier to prevent the spread of infectious disease. This can include separate clothing, PPE and equipment for use in infected areas, limits on the people who are allowed access to the area and isolation of infected animals.

PPE used in barrier nursing can include face masks, gloves, disposable coveralls or gowns and goggles.

- Hygiene is vital in barrier nursing; all people working in infected areas must ensure they are washing their hands rigorously with soap and water as well as using hand sanitisers before and after contact with the patient.

- Proper disposal of contaminated materials such as gloves, gowns, substrate and animal waste must be carried out in accordance with regulations.

- Many facilities where the risk of airborne diseases is high have rooms equipped with negative pressure ventilation to prevent pathogens from escaping into other areas.

Testing, inoculation and vaccination

A lot of diseases and pests do not show symptoms until they are very well established. **Carrying out regular tests,** for example faecal egg counts, can highlight where treatment needs to be administered before the problem spreads.

In addition to testing, a vaccination and worming schedule can prevent outbreaks and reduce the severity when they do happen.

Vaccinations are used to create immunity to specific diseases. A vaccination contains weakened, dead or parts of a pathogen, such as a virus or bacteria that exposes the immune system to the pathogen in a safe way. This allows the immune system to prepare antibodies that are ready in the event of infection.

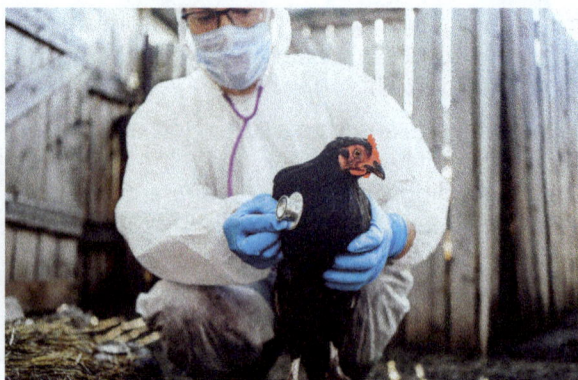

A vet carrying out a health check whilst wearing PPE to ensure biosecurity is maintained

Inoculations are similar to vaccinations but they involve the introduction of a pathogen or antigen (something foreign to the body which the immune system reacts to) into the body. This causes a mild infection which the immune system reacts to and fights.

Remember – prevention is better than cure!

Elimination of vectors

Vectors are animals, pests, parasites, people or objects that can carry disease.

Animals new to the collection should be quarantined to ensure they are free from disease before entering the general population.

Pests such as mice, rats or flies should be controlled as they can carry disease.

Staff should be trained to ensure high standards of hygiene for themselves as well as enclosures and equipment, and visitors should be restricted to areas where they cannot put animal health at risk.

Prophylactic treatment can prevent parasite infections and reduce the risk of disease.

Education and training

One of the most important aspects of biosecurity is training for everyone working with animals.

Hygiene, including handwashing and keeping a clean working environment, will help to reduce the introduction and spread of disease but it is not just staff who need to wash their hands when handling or working with animals. Collections which encourage members of the public to touch or hand-feed animals must employ measures to ensure those people have easy access to hand washing facilities, with clear signage explaining the importance of doing so.

Events where children are in contact with animals must pay special attention to hand washing. Children are at particular risk because they often put their hands to their face or their fingers in their mouth.

> **Important terms!**
>
> Vector: An organism that carries an infectious disease but is not affected by it. For example a mosquito carrying malaria.
>
> Prophylactic: Preventative treatments, such as vaccinations or worming, that are used to prevent or reduce the severity of disease.
>
> Passport: An animal passport is a legal document that identifies an animal and can be used to track its movement. All horses and cattle in the UK must have a passport.
>
> Efficacy: How effective something is.
>
> Faecal egg counts: A test to see how many parasite eggs, if any, are found in an animal's faeces.

Implications of the lack of biosecurity

Biosecurity has a lot of implications beyond animal and human health. If disease is allowed to spread out of the collection, there is a risk to the wider community in terms of both human, domestic and wild animals. If a company is found to be guilty of releasing disease into the environment their business reputation will suffer. Negative publicity for something as serious as a biosecurity breach can lead to customers losing faith in the organisation and taking their business elsewhere. This will obviously have economic impacts.

Biosecurity breaches can put staff and visitors at risk. Staff wellbeing should be a top priority for employers no matter what industry they are in, as wellbeing is intrinsically linked to satisfaction and, therefore, performance.

Of course, biosecurity breaches will put animal health and welfare at risk. Disease outbreaks can impact large numbers of animals in the collection, leading to loss of stock and therefore profit.

Biosecurity risk factors

There are several risk factors that must be considered when planning biosecurity measures. Key factors relating to environment, animal husbandry and animal health must all be planned for to avoid issues arising.

Environmental factors

- **Humidity** levels must be carefully monitored because excessive moisture can create conditions that promote the growth of harmful bacteria, fungi, or moulds.

- Proper **ventilation** is equally important to ensure a constant supply of fresh air, reduce the buildup of ammonia or other harmful gases, and minimise the risk of airborne pathogens spreading.

- **Temperature control** is another key element, as extreme heat or cold can stress animals and weaken their immune systems, making them more susceptible to disease.

- The **materials used in constructing enclosures** or facilities also impact biosecurity: materials that are difficult to clean or prone to degradation can harbour pathogens.

- **Resources such as tools and feeding equipment** should be cleaned regularly and maintained in good condition to prevent contamination.

- The **presence of wildlife** around animal enclosures poses another risk, as wild animals can introduce diseases or act as carriers for pests and parasites.

- **Effective drainage systems** are essential to prevent standing water, which can attract pests and become a breeding ground for bacteria.

- **Organic pollution,** such as the accumulation of waste or decaying matter, must be managed to prevent the spread of harmful microorganisms.

Animal health

Animal health considerations are another essential aspect of biosecurity.

- The **carrier status of animals,** particularly during disease incubation periods, poses a significant risk. Animals may appear healthy but still carry and spread infectious agents to others.

- **Vaccination** is a key preventive measure, and it is important to ensure that animals are up-to-date with their vaccinations to reduce susceptibility to disease.

- **Stocking density,** the number of animals housed in a given space, must be managed carefully to avoid overcrowding, which can increase stress levels, reduce air quality, and facilitate the rapid spread of disease.

- **Vector control methods,** such as managing insect populations or other disease-carrying organisms, are crucial in reducing the risk of diseases being transmitted within or between animal populations.

Animal husbandry

Husbandry practices also have a direct impact on biosecurity.

- **Proper food and water provision is vital,** as contaminated feed or water can introduce harmful pathogens into the system.

- **Hygiene standards must be maintained** across all aspects of animal care. This includes ensuring that people involved in animal care adhere to strict hygiene protocols, such as using personal protective equipment (PPE) and washing hands thoroughly to minimise the risk of disease transmission.

- **Animals and their accommodations should be cleaned and disinfected regularly** to remove potential pathogens and maintain a healthy environment.

Biosecurity under current legislation

Due to the importance of biosecurity, several pieces of legislation have key points that stipulate the requirements that owners and staff caring for animals must adhere to. Below is a summary of specific points relating to biosecurity.

Animal Welfare Act 2006

> **Important terms!**
>
> Collection Manager: The person with overall responsibility for animals and staff within a collection. They will often be involved with writing policies, carrying out risk assessments and delivering training.

Duty of care

Legislation makes it the responsibility of the collection manager to ensure that animals are protected from disease and injury. This includes implementing biosecurity measures to prevent the introduction and spread of disease.

Health and treatment

Those responsible for the care of animals in the collection must provide appropriate treatment for ill or injured animals. This includes maintaining clean environments and preventing contact with potentially infectious animals. Animals that are showing signs of a potentially infectious disease should be separated from the main population to reduce the spread of disease.

Environment

People working with animals must maintain hygienic conditions and provide proper housing to support animal health and prevent disease outbreaks.

Hygiene is vital not just for animal welfare but for human health and safety as well. Ensuring high standards and protocols are in place will reduce the risk of infection and prevent outbreaks from becoming serious.

Control of Substances Hazardous to Health (COSHH) 2002

Risk assessment

Collections must conduct risk assessments for hazardous substances, including biological agents, to identify and mitigate risks to health.

Common hazardous substances in animal collections include medications and cleaning products.

Risk assessments attempt to quantify how hazardous something is, along with how likely it is to cause harm. For more information on risk assessments see Section 1.

Biological agents under COSHH are any disease-causing agents, for example bacteria, fungus, virus and prions. COSHH rules for biological agents apply specifically to laboratories, colleges and universities that are working with these pathogens to ensure staff are safe. Protocols and safeguards must be in place to ensure pathogens are not released from the laboratory.

Control measures

Once risk has been assessed, control measures are put in place to minimise that risk. Collections must implement appropriate controls to prevent exposure to hazardous substances. Typical control measures include:

1. Disinfection to remove disease-causing pathogens

2. Personal protective equipment (PPE) to stop hazardous substances from coming into contact with the skin

3. Training to ensure all staff know the dangers and how to avoid them

Disinfecting a farm wearing full PPE

In the majority of animal collections, COSHH control measures will be limited to medications and cleaning products that are used on a daily basis. These projects can be very harmful to animals (and humans) so it is important that risk assessments identify the hazards and control measures are put in place to minimise the risks.

Training and information

All staff working in environments where biosecurity is important should be trained to ensure they know the risk that are present. This training will include:

1. How to identify possible signs of infection

2. Proper use and storage of chemicals

3. Actions to take to reduce the spread of disease, such as foot dips and the use of specific hygiene practices.

Signs must be displayed where appropriate to remind staff about the risks and how to reduce them.

Zoonoses Order (Amendment) (England) 2021

Notification and reporting

Certain zoonotic diseases must be reported to the local authority to facilitate monitoring and control. Once a zoonotic disease has been identified and reported, organisations such as Defra and APHA will then help to put relevant control measures in place. This will include:

1. Advising on specific measures that need to be taken.

2. Implementing protocols to reduce the number of people in the area, for instance closing public footpaths temporarily.

Control measures

Organisations must implement specific control measures to manage and reduce the risk of zoonotic disease transmission between animals and humans. These preventative control measures are designed to reduce the risk of infection occurring within the animal collection.

If these preventative measures are unsuccessful and infection has been discovered, addition control measures will then be necessary to reduce the spread of the disease.

Control measures will include everything from:

1. Training staff

2. Using signs to educate staff and visitors and encourage compliance

3. Use of PPE

4. Hygiene practices

5. Isolation and quarantine

6. In extreme cases, culling animals that are infected.

45

Surveillance and monitoring

Organisations must conduct surveillance and monitoring of zoonotic diseases to detect and respond to outbreaks promptly. This will include:

1. Regular health checking of animals
2. Vaccination and worming schedules
3. Quarantining new animals
4. Isolating animals that are showing signs of infection
5. Blood and faecal sampling

Animal Health Act 2002

Organisations must implement measures to prevent, control and eradicate animal diseases, including quarantine, vaccination and movement restrictions.

This piece of legislation means that it is not just a good idea to implement these measures but it is a legal requirement to do so. This ensures that collections that have not put protocols in place can be held accountable.

Biosecurity standards

Collections must adhere to biosecurity standards and protocols to prevent the introduction and spread of infectious diseases. Organisations that are found to be the cause of disease outbreaks will face severe penalties.

Inspections and compliance

Organisations must allow inspections by authorities to ensure compliance with animal health regulations and biosecurity practices. Local authorities will conduct inspections to ensure that the animal collection is complying with animal legislation relevant to the type of establishment.

Convention on International Trade in Endangered Species of Wild Fauna and Flora (CITES)

CITES was established in 1973 due to concerns that wild animal and plant populations were at risk of international trade. Countries joined this international organisation in an attempt to save species from extinction.

Regulation of trade

CITES ensures that international trade in endangered species does not threaten their survival. It includes measures to prevent the spread of disease through trade. There are three categories within CITES that plant and animal species are added to depending on their current status:

1. **Appendix I** within CITES includes species that are threatened with extinction. Trade in these species is forbidden except in exceptional circumstances.

2. **Appendix II** includes species that are not necessarily currently threatened with extinction, but countries need to control trade in wild specimens to prevent the threat of extinction in the future.

3. **Appendix III** includes species that are protected in at least one country. These are often species that have become very popular in the pet trade, often due to media influences.

Permits and documentation

Individuals and organisations require permits and proper documentation for the import, export and re-export of listed species to ensure compliance with biosecurity measures. Animals that are imported or exported must undergo strict quarantine to ensure the risk of disease is minimised.

Monitoring and reporting

CITES monitors all trade in listed species to ensure that individual species are not being exploited. They monitor and report trade activities to ensure that biosecurity risks are managed and that species are not exposed to new diseases through international movement.

Recap Questions

1. What is the purpose of biosecurity? Give at least three examples in your answer.

2. What is an invasive species? Why are they a problem?

3. How can public awareness of biosecurity be improved? Why is this important?

4. What role do Defra and APHA have in managing biosecurity in the UK?

5. What are the implications of breaching biosecurity measures?

6. How does culling infected animals impact farmers?

7. Explain three factors that can affect biosecurity either positively or negatively.

8. Why does the illegal pet trade cause significant problems for national biosecurity?

9. What is the difference between a zoonotic, notifiable and anthroponotic disease? Give examples of each in your answer.

10. What is the role of Defra and APHA is managing disease outbreaks?

11. How can implementing restriction zones around infected sites reduce the spread of disease?

12. Explain five different control measures that can be used to maintain biosecurity. Discuss their effectiveness in different types of collections.

13. Why is education on biosecurity not just important for staff working in animal collections?

14. How can testing, inoculation and vaccinations help with biosecurity?

15. Discuss how vectors can be controlled. How does this help with biosecurity?

16. Why is ventilation a potential risk factor for biosecurity?

17. What is meant by the term 'carrier status of animals'?

18. Why is hygiene important in an animal collection?

19. Name three pieces of legislation that include requirements for biosecurity.

20. What is the role of Defra and APHA in biosecurity?

21. What control measures are required under legislation to prevent disease outbreaks?

Real world example

In 2001, an outbreak of Foot-and-Mouth Disease (FMD) occurred in the South of England. Foot-and-Mouth is a highly contagious virus that affects cloven-hooved animals such as cattle, pigs, sheep and goats.

Once the outbreak was reported, animal movement restrictions were put in place by Defra and executed by APHA. This put an immediate ban on the movement of animals to prevent the spread of the disease between farms.

In addition to movement restrictions, quarantine zones were established around infected premises, limiting access to people and traffic. Strict disinfection protocols were put in place for people, equipment and vehicles entering or exiting any of the quarantine areas.

To prevent the spread of disease between farms, and ultimately prevent infected animals from entering the human food chain, all cloven hooved animals on infected farms were culled.

Extensive testing and monitoring of livestock was carried out under Defra's orders to identify new cases as soon as they occurred, so that culling could take place before it had spread.

The media was involved in reporting the situation to ensure that farmers and the general public across the UK were aware of the risks and the symptoms to look out for.

Following these actions, the disease outbreak ended in January 2002, 11 months after the outbreak was initially discovered. Once the UK had reported no cases of FMD for three months, the country was declared free of the disease. This case highlights the importance of biosecurity and how effective management can reduce the risk to human and animal health.

Practice Questions

1. Evaluate the importance of legislation in relation to biosecurity. Your answer should include reference to specific pieces of legislation. (6 marks)

2. What are the potential financial and environmental consequences of poor biosecurity in an animal care facility? (5 marks)

3. A disease has broken out in the animal collection you are working in. Describe the steps that the organisation must now take in order to contain the disease. (8 marks)

4. When bringing new animals into an animal collection, what biosecurity measures should be in place and why? What could the consequences be of not following these measures? (6 marks)

5. Assess the role of biosecurity in conservation programmes. (5 marks)

6. One of the main roles of biosecurity is maintaining human health and safety. Explain how adequate staff training with regards to biosecurity can ensure staff safety. (5 marks)

7. Many animal collections that are open to the public face difficulties with biosecurity. Explain the problems an urban farm which offers visitors contact with the animals may face in maintaining biosecurity. What can they do to overcome these issues? (8 marks)

4.1 Principles of supply chains in the animal management sector

What is a supply chain?

A supply chain is a network of businesses that are involved in getting a product or service to a consumer/customer. It starts with raw materials and ends with the final product or service.

Different businesses are often involved in each stage of a supply chain.

What makes up a supply chain?

- **Raw Materials** – the initial materials used to make the product that come from suppliers

- **Transportation** - the movement of materials from one stage of the chain to the next

- **Production** – the process of turning the raw materials into the product

- **Distribution** - the movement of the final product from the manufacturer to the consumers

- **Delivery** - the final product arrives with the consumer; retail or customers

Animal Management Sector Example: Dog Bed Supply Chain

- **Raw Materials** – Stuffing, fabric for cover, thread.

- **Transportation** - Sent from manufacturers to company making beds.

- **Production** – Company makes beds.

- **Distribution** - Beds sent to shops or direct to customers.

- **Delivery** - Beds arrive in shops or with customers.

Impact of legislation and regulation

Various pieces of **legislation and regulations** impact on the animal management supply chain.

Convention on International Trade in Endangered Species of Wild Fauna and Flora (CITES)

- This is an international agreement between worldwide governments.

- Its aim is to ensure that **international trade** in specimens of wild animals and plants **does not threaten the survival** of the species. It covers around 5,000 animal species.

- It puts controls on the international trade of certain species. **Licenses** are needed to import, export and re-export of the species. It mainly affects the pet trade business, and particularly the exotic pet trade business.

- It also covers goods made from animal products.

Example: Many parrots species are covered by CITES

The Animal Welfare (Licensing of Activities Involving Animals) (England) Regulations 2018

- These regulations establish **licensing requirements and welfare standards** for various activities involving animals, ensuring businesses in these supply chains maintain high welfare standards.

- Examples: ethical sourcing of animals, welfare friendly handling and transportation of animals.

Animal Welfare Act 2006

- This law sets out the **five welfare needs** of animals.
- Businesses involved in any step of the supply chain must adhere to these as part of their duty of care. They must also ensure products for animals meet welfare standards.

The Welfare of Animals (Transport) (England) Order 2006

- This order establishes **rules for the transportation of live animals** to ensure their welfare during transit.
- The transportation of live animals is a critical part of the supply chain for industries such as agriculture, food production and live animal trade.
- The transport stage of a supply chain is a vital process within the animal trade.
- It protects the animals but also the integrity, traceability and sustainability of the supply chain.
- The welfare-focused approach supports sustainable supply chains by emphasising ethical treatment and reducing waste due to injuries or unfit animals.

Veterinary Surgeons Act 1966

- This Act ensures professional, ethical and welfare-focused veterinary care which upholds the health and productivity of animals.
- **Only qualified and registered veterinary surgeons can perform certain procedures** on animals. This ensures professional standards are met.
- Healthy animals are essential for productivity and quality in supply chains involving livestock, poultry and aquaculture.
- For supply chains involving the international trade of live animals or animal products, vets can verify the fitness of the animals for travel and trade.

Loading sheep into a trailer

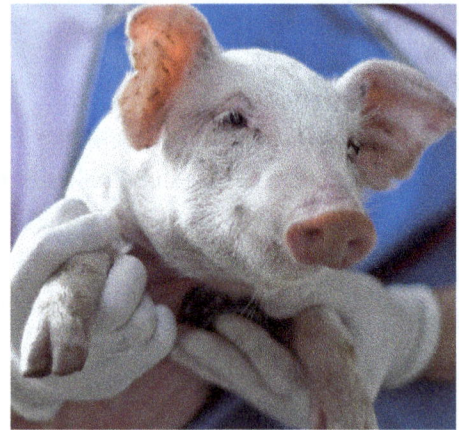

Pig examination with stethoscope

Importance of efficiency and interdependency

Efficiency is working well in an organised way without wasting time or energy.

Interdependency is the relying of two or more things on each other.

They are both important to a supply chain:

- to reduce customer demand
- to reduce costs within the chain
- to reduce waste within the chain
- to ensure seamless flow within the chain
- for fast deliveries
- to ensure reliability and flexibility
- to ensure profitability.

Suppliers, distributors and customers

- Suppliers need to provide a timely flow of raw materials to the manufacturers.
- Distributors provide the products to the retail stores and customers from the manufacturers.
- Customers are the final goal of the supply chain, whether directly ordered, for example via online ordering, or indirectly via retail stores.

- **Each stage relies on the stage before** performing in a timely and organised manner to meet customer demand and to remain competitive within the sector.
- **If one part of the chain fails it will impact the rest of the chain.** Equally, a decrease in customer demand will cause issues earlier in the chain.

Supply chain assurance

This is the process of ensuring that **all elements of a supply chain operate effectively, efficiently and safely** to deliver quality compliant products.

It supports both the businesses in the supply chain and the customers.

Ethics

Ethics are the **moral principles, policies and values** that govern the way businesses and individuals conduct their activities.

It is more than meeting legal requirements.

It is a code of conduct that should drive all business activities and employee behaviour, and helps build trust between a business, its suppliers and its customers.

Ethics matter in supply chains because they:

1. **Builds trust** and reputation.

2. **Ensures compliance** with laws and regulations, reducing the risk of legal penalties and sanctions.

3. Promotes **fair treatment** of workers.

4. Encourages **sustainability** and a fair global economy.

5. Enhances risk management to **reduce the risk** of scandals, fraud or supply chain disruptions.

6. Provides **transparency and accountability** in the supply chain.

7. Aligns with modern **consumer expectations** for responsible sourcing and production.

8. Supports economic and social development by **supporting local economies**.

Parts of a supply chain

The key elements are:

- **Risk Management**: Identifying, assessing, and mitigating risks (e.g. disruptions, quality issues, or cybersecurity threats) to ensure supply chain continuity.

- **Quality Assurance**: Maintaining consistent quality standards across all suppliers and processes to meet customer expectations.

- **Compliance**: Ensuring all supply chain activities adhere to legal, ethical, and regulatory requirements.

- **Transparency**: Providing visibility into processes, sourcing, and logistics to build trust and enable informed decision-making.

- **Collaboration**: Fostering strong partnerships and communication among stakeholders to enhance coordination and problem-solving.

- **Resilience**: Developing strategies and contingencies to withstand disruptions and recover quickly.

- **Performance Monitoring**: Regularly evaluating suppliers, processes, and systems to ensure efficiency and alignment with objectives.

- **Sustainability**: Integrating environmentally friendly and socially responsible practices into supply chain operations.

Importance of ethical sourcing of stock and services

Ethical sourcing refers to the process of obtaining materials, products or services in a way that respects principles, such as **fairness**, **environmental sustainability** and **social responsibility**.

It ensures that every stage of sourcing **upholds legal and moral standards**, benefitting workers, communities and the environment.

Ethical sourcing is a **commitment to creating a positive impact** on people, the planet and the future of business.

Important elements of ethical sourcing

1. Fair labour practices – workers treated fairly

2. Environmental sustainability – minimising environmental harm

3. Transparency – clear visibility into business practices

4. Compliance with laws and standards – local and international laws and regulations

5. Community impact – supporting local economies

Benefits of ethical sourcing

1. Improved brand reputation – building trust

2. Consumer loyalty – return customers

3. Risk mitigation – reduces risks within chain, legal issues

4. Sustainable growth – long-term relationships and resilient chains

5. Positive global impact – working conditions, conservation and well-being

Traceability in supply chains

- **Traceability** means tracking the movement of products, components or materials throughout the supply chain, from origin to final destination.

- Involves **documenting and monitoring** every step of the process.

- Ensuries product quality and safety, transparency, accountability and efficiency.

The benefits of traceability

- **Enhances consumer trust** about the origins and handling of products, which is increasingly valued by consumers.

- **Reduces exposure to fraud**, counterfeit goods and unethical practices.

- **Enhances inventory management** by providing real-time visibility of stock levels and locations.

- Encourages suppliers to maintain high standards, knowing their processes are monitored.

- Provides valuable data for optimising supply chain operations and predicting future trends.

Animal Management stock and services examples

Ethical sourcing ensures that the production, processing and supply of the items listed below **do not harm ecosystems, exploit labour or compromise animal health**.

Feeds (raw, live)

Animal feed is an important stock item for any animal management sector business.

It may be that the feed is required for the animals kept by that business (e.g. zoos). Alternatively the business may supply feed to its customers (e.g. pet shops).

- The feed is either in its **natural raw state**, such as wheat and maize, or in a **processed state**, such as dog and cat kibble. It can also be in a **live state** such as crickets and meal worms.

Clinical supplies

Clinical supplies are items that are used in veterinary practices, treatment and first aid by owners and keepers.

Examples include surgical equipment, bandages and dressings, gloves, cotton wool, PPE, needles, sharps bins, disinfectant wipes and thermometers.

Bedding

Bedding materials vary and include:

- substrates within an animal enclosure, such as wood shavings, straw, wood pellets, bark, sand or rubber matting

- fabric bedding such as dog and cat beds and blankets.

Animal food

> **Remember!**
> It is illegal to feed live vertebrate prey to another animal in the UK.

Equipment including PPE and animal housing

PPE and animal housing

General equipment for the animal management sector is a vast category and covers anything required for keeping animals outside of feed, bedding and clinical supplies. It includes housing (from large to small animals), handling equipment and husbandry materials such as brushes, PPE, buckets and disinfectants. It also includes toys and enrichment for animals, which is a big part of the pet industry for owners.

Deadstock/cadaver disposal services

A cadaver is a dead body. Services in the animal management industry cover the **disposal of dead animals**. This ranges from deadstock in agriculture and animal collections to those in pet shops and animals kept as pets.

- Regulations exist to ensure safe disposal services exist for agricultural animals and veterinary practises.
- Pet cremation is a common service provided for pet owners and is a regulated industry.

Animal stock and the importance of welfare standards and ethical sourcing

Animal welfare standards ensure animal stock is treated humanely and prevent animal suffering due to poor welfare.

Ethical Sourcing ensures animal stock is obtained and managed in a way that respects their welfare and natural environments. It discourages exploitation, illegal trade or abusive practices. For example:

- **Wildlife**: Animals captured from the wild should adhere to strict conservation laws.
- **Pets**: Breeders and sellers should prioritise animal health and avoiding mass breeding (e.g. puppy farms).

Pet trade

The **pet trade** involves breeding, selling and distributing animals for domestic companionship.

It includes common pets like cats, dogs, fish, birds, reptiles and exotic animals.

Issues can include over-breeding and poor care, wild-caught species facing stress and population decline, disease spread due to poor animal handling and illegal trade harming endangered species.

Ethical sourcing (reputable breeders) and welfare standards can reduce issues within the pet trade.

Dealers

Pet dealers are individuals or businesses that sell animals to the public, breeders or retailers.

Some dealers engage in unethical practices, such as:

- Selling wild-caught animals under poor conditions.
- Failing to meet welfare regulations (e.g. overcrowded or unsanitary cages).
- Misrepresenting the origins or health of the animals.

Responsible dealers are those who prioritise animal welfare, provide proper care instructions and ensure traceability.

Reputable breeders

Reputable breeders are individuals or organisations that ethically breed animals, prioritising health, welfare and genetic diversity. They improve animal welfare and support sustainable breeding.

Key traits of reputable breeders:

- Follow welfare standards for food, housing, medical care and enrichment.
- Avoid over-breeding and genetic issues.
- Provide full health records, genetic testing and vaccination.
- Educate buyers about care needs and screen prospective owners.

Wild-caught

Wild-caught animals are taken directly from their natural habitat for the pet trade or other purposes.

Concerns include:

- High stress and mortality rates during capture and transport.
- Disruption to ecosystems and declining wild populations.
- Risk of introducing diseases into new environments.

International laws like **CITES** (**Convention on International Trade in Endangered Species**) regulate wild animal trade to protect endangered species.

Captive-bred or captive-hatched animals reduce harm to wild populations and can be alternative sources.

Captive-hatched

Captive-hatched animals are born in captivity from eggs that were collected from the wild.

This reduces stress compared to capturing adult animals and can support conservation efforts when well-regulated.

Regulation and ethical sourcing are critical. For example, if overexploited, egg collection can still harm wild populations.

Captive breed

Captive breed animals are bred and raised entirely in captivity.

The benefits of captive breeding include:

- Reduction of pressure on wild populations.
- Animals are typically healthier, used to human care and less prone to disease.
- Supports ethical pet trade when done responsibly.

Poor captive breeding practices, like inbreeding or mass production, can lead to genetic and health issues.

Traceability

Traceability is the ability to track an animal's origin, ownership history and health records throughout its life.

- It ensures animals are ethically sourced and legally traded.

Seahorse are commonly bred in captivity

- It allows for the identification of breeders, dealers or importers who violate welfare standards.
- Supports conservation efforts by monitoring wild populations and reducing illegal trade.

The following tools help with traceability: microchipping, genetic testing, breeding records and certification systems.

Services in animal management supply chain

Live feed distributors

Live feed distrbutors are suppliers that specialise in providing live organisms used as feed for animals.

- Examples include for aquaculture, pet care and zoological facilities.

- They ensure that live feed is of high quality, nutritionally appropriate, and delivered in a way that maintains the health and welfare of the feed organisms.

- Types of live feed include: brine shrimp, crickets, mealworms, waxworms, earthworms and grubs.

Edible insects

Wholesalers

- A **wholesaler** is a company or individual that buys products in large quantities from manufacturers or producers and sells them in smaller quantities to retailers or other businesses.

- Wholesalers operate between manufacturers and retailers. They make money by buying in bulk at a lower price and then adding a profit before selling the products on.

- Animal wholesalers sell a variety of products and animals in bulk to businesses rather than directly to individual consumers. Their primary customers are pet stores, farms, zoos, aquariums, research institutions, veterinary clinics and other organisations that require animals or animal-related products.

- Examples of their products include; live animals, animal feed, animal supplies and equipment.

Fallen stock services

- **Dead livestock** are also known as fallen stock. In England, horses can be buried or dealt with as fallen stock. Pets can be buried or cremated.

- Farmers must ensure the safe, legal collection and disposal of fallen stock and cover the associated costs.

- The government issues guidance on the safe and legal disposal of fallen stock.

- Disposal options are using the National Fallen Stock Company (NFSCo) for collection and disposal or arranging collection by an approved transporter for disposal at fallen stock servicers: knackers, hunt kennels, maggot farms, incinerators or rendering plants.

Crematoriums

Pet crematoriums are increasing in popularity. They provide individual or communal cremations for pets, with collections from homes or veterinary practices, and ashes returned to the owners or communally scattered at a suitable location.

Pet crematoriums are accredited by the APHA (Animal and Plant Health Agency) to make sure the standards set out by the government are met.

Boarding kennels

Boarding kennels offer care for dogs when their owners are away from home.

Boarding catteries provide the same service for cats.

Some boarding facilities offer accommodation for other commonly-kept pets, such as rabbits and guinea pigs.

A pet urn

Clinical / laboratory services

- These services analyses samples to help **diagnose, treat and manage animals**.

- Common samples are blood, urine, tissues and other bodily fluids.

- They identify abnormalities, pathogens and changes in biochemical markers.

- They provide this service to **veterinary practices and hospitals**.

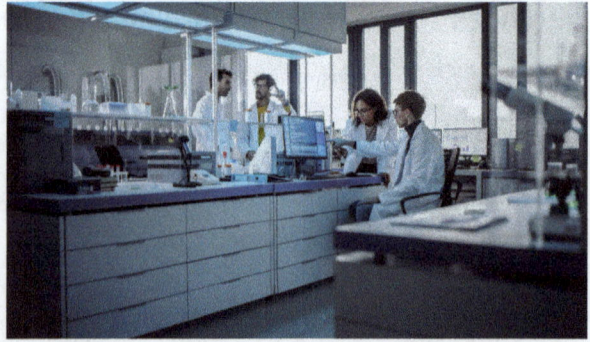
A scientific laboratory

Veterinary services

- **Veterinary serices** include the examination, diagnosis and treatment of animal patients, administration of vaccines, diagnostics, imaging, surgery, laboratory testing and provision of hospitalisation and emergency treatment.

- They are provided by registered qualified veterinary surgeons and associated staff.

Supply of animals

- **Animals are supplied for several reasons**, including: food, pet trade, show trade, sport, agriculture, working animals, hobbies, zoological collections and breeding.

- Regulations and legislation must be followed.

- **Animal welfare is a priority**, covering animal wellbeing and safe and sustainable production. The **Animal Welfare Act 2006** requires proper living conditions, nutrition, medical care and preventing cruelty and neglect.

- Illegal wildlife trade and non-ethical sourcing must be avoided.

Supply chain sequences and operations

Supply chains, whether for animals, goods or services, can be sequenced and operated in various ways.

Key methods and their typical operations examples are:

- **Direct Supply Chain: Producer to Consumer**: This is a straightforward supply chain where the producer directly sells to the consumer. For instance, a local farmer selling eggs at a farmers' market.

- **Extended Supply Chain: Producer to Distributor to Retailer to Consumer**: This is a more complex chain where goods pass through multiple intermediaries. For example, livestock raised by a farmer, sold to a slaughterhouse, then to a distributor, and finally to a supermarket for consumer purchase.

- **Supply Chain Network: Multiple Suppliers and Distributors**: This involves multiple suppliers and distributors working together to supply a single product. For example, a pet store might source pets from several breeders, food from multiple manufacturers, and accessories from different suppliers.

- **Lean Supply Chain: Just-in-Time (JIT)**: This method aims to reduce waste by receiving goods only as they are needed in the production process. For example, a zoo might order food for its animals on a weekly basis to ensure freshness and reduce storage costs.

- **Agile Supply Chain: Flexible and Responsive**: Designed to be highly flexible to respond quickly to changes in demand or supply conditions. For instance, a wildlife conservation project might need to quickly adapt its supply chain to respond to emergencies, such as a natural disaster affecting animal habitats.

- **Digital Supply Chain: Technology-Driven**: Utilises advanced technologies such as IoT (Internet of Things – for example smart devices), blockchain (decentralised system for recording data across a network of computers), and AI to improve efficiency, transparency and collaboration. For example, using RFID tags to track the movement of cattle from farm to slaughterhouse to ensure traceability.

- **Circular Supply Chain: Sustainable and Regenerative**: Focuses on recycling and reusing resources to minimise waste. For instance, using animal waste as fertiliser for crops that are then used to feed animals, creating a closed-loop system.

Types of suppliers

Primary

Primary suppliers are those directly involved in the production of goods or delivery of core services. They provide the main materials, components or services that have a significant impact on the quality and functionality of the final product or service.

Secondary

Secondary suppliers provide components, materials or services that support the production process of primary suppliers.

Tertiary

Tertiary suppliers are businesses that provide goods or services to secondary suppliers

Animal Management example: you have primary suppliers providing animals, secondary suppliers providing the necessary supplies such as feed and bedding for the animals to the primary suppliers, and tertiary suppliers offering raw materials for the food and bedding.

• •

Implications of failing to meet supply chain demands

Quantity

If a product or service is **not supplied in enough numbers**, then a business may face lost sales and reduced revenue due to shortages in supplies.

There may also be increased costs for the business due to the need for faster shipping, or overtime for workers or alternative suppliers to make up any shortfall in numbers.

Quality

Poor quality in the supply chain can have serious effects, leading to numerous negative consequences across various stages. Some major impacts include returns and replacements, product recalls, negative reviews, production delays, legal issues and fines, loss of market share and increased waste.

Safety – human and animal

Animals can be at risk of **poor welfare due to failings in the supply chain**. It could be linked to their housing, husbandry, food or health, depending on the supply chain involved. Those responsible for animals have a duty of care to provide the five animal needs and to not put safety at risk.

Humans can be put at risk physically and mentally by a failing supply chain.

• Stress-related illnesses can result from pressures due to supply chain issues.

• The supply chain can also impact human safety if there is a lack of PPE provided for animal husbandry, for example, or poor-quality handling equipment leading to injuries.

Staff wellbeing

• Business owners can have increased stress and anxiety due to being unable to sustain their businesses.

• Job losses can result from a failing supply chain which can impact on staff wellbeing for those who are made redundant and increased pressure on those left in the business.

• There can be increased stress and anxiety due to the inability to care for their animals properly if products are not available.

Business model

A business model is a plan that outlines how a business operates, makes money and sustains itself in the market. It should identify the key components of a business and the interactions to achieve its goals.

The impact of failing to meet supply chain demands can impact through every aspect of a business model.

Examples of areas affected are:

• loss of revenue and profitability

• lost sales

• increased costs

• poor customer relationships and trust

• negative reputation

• loss of market position and market share reduction

• cash flow problems

• low creditworthiness.

Persistent supply chain failures may force a re-evaluation of the entire business model, possibly leading to strategic changes or restructuring.

Ensuring **robust supply chain management** and ha**ving contingency plans** in place are crucial for maintaining **business continuity** and **long-term success**.

External influences and effects

External influences (also called external factors) are things that happen outside of the business that can impact on the business in a positive or negative way.

1. They are **beyond the control** of the business.

2. The business **must react to them** to remain successful.

There are six key external factors affecting businesses:

*   **Political** – driven by government and politics, for example, new government policies, change in government or world politics, for example trade agreements.

*   **Economic** – driven by the economy, for example, inflation (prices of raw materials), unemployment, interest rates.

*   **Socio-cultural** – driven the way societies are organised and traditions, beliefs and practices, for example, changes in taste or fashion or new trends such as raw pet diets and clothing for dogs.

*   **Technological** – driven by advances in technology, for example, online shopping, automated procedures and factories, artificial intelligence.

*   **Legislative** – driven by laws, for example, new laws or regulations being introduced or amendments to existing laws which can restrict goods and services.

*   **Environmental** – driven by environmental issues and sustainability, for example, weather, climate change, recycling, carbon footprints.

Political	Economic	Social	Technological	Legal	Environmental

P E S T L E

Customer base

A customer base is the group of customers who regularly buy the products or services that a business offers. It can also be called a **client base**.

They are the primary source of revenue for a business. These customers are the core audience that the business serves. They often represent the most loyal and frequent buyers, making them critical for the company's revenue and long-term success.

The **target market** is the potential customers that a company aims to attract with its marketing efforts. It represents the ideal audience for a business's offerings.

Target market

The larger the customer base (the number of customers) the greater the revenue.

In a pet shop the regular customers are the customer base. The shop owners can target the broader area and online customers (the target market) to attract more business.

Types and characteristics

The customer base can be divided up as the following types of customers:

Type	Characteristics	Animal Example
Loyal	These are regular customers who frequently purchase goods or services. Will recommend business to others.	Regular clients at a grooming parlour.
Potential	The customers in the target market.	People with pets who have moved into a new housing development near a pet store.
Occasional	Irregular customers who occasionally purchase goods or services, or only purchase on special occasions.	Visitors to a zoo who may go once, or only for a special treat such as a birthday or anniversary.
New	Customers who have recently started to purchase goods or services.	Owner using a hydrotherapy pool for their dog as it is starting to get older.
High value	Customers who make significant or high-volume purchases.	Farmers who may purchase feed or bedding materials in bulk.
Niche	Customers with very specific needs or preferences.	Hobbyists who are looking for particular colourations or patterns of snake and lizard species.
Digital	Customers who will only purchase goods or services online.	Customers of a bespoke pet identity tag company who supply tags and collars.
Budget conscious	Customers who prioritise affordability over other factors.	Rabbit owners who are looking for affordable hutches.
Former	Customers who are no longer using the business but who could be encouraged to return.	Bereaved pet owners who may return if they get new pets in the future.

Market segments

Market segments are distinct groups of customers within a broader market, categorised based on shared characteristics, behaviours or needs.

They allow businesses to **segment**. This means tailoring their products, services and marketing efforts to meet the specific preferences of each group more effectively.

Here are some examples:

*   Pet food companies segment their products – not just for different species but for other characteristics, such as age, activity level, health, skin/coat condition, and for ingredients (such as 'grain free').
*   Pet retail stores can segment their pet accessories into sustainable eco-friendly, high-end luxury and affordable categories to appeal to broader markets.

Dog food at a pet shop

Competitor analysis

Competitor analysis is the process of identifying and researching the competitors of a business.

*   It involves looking at the strengths and weaknesses of the competitors and comparing them with the business.
*   It can help to identify opportunities for the business and help the business to grow.
*   It can highlight the differences between the business and its competitors.

*   By understanding what competitors are doing, the business can refine its goods or services, adjust pricing and create marketing campaigns to better meet customer needs and gain a competitive edge.

Direct competitors are those who are offering similar goods or services to the same target market.

Indirect competitors are those who are offering alternative goods or services that meet the same needs.

Animal Example: For a pet food company:

- **Direct Competitors**: Other pet food brands selling to the same target market (e.g. grain-free pet food).

- **Indirect Competitors**: Home-cooked pet meal solutions or raw food suppliers.

Types of procurement

Procurement is the obtaining or purchasing of products or services for a business.

- **Direct procurement** relates to the main products or services of a business and impacts on profit.

- **Indirect procurement** is the expense of day-to-day operations of the business and impacts on efficiency.

Bulk purchasing and direct purchasing are two different procurement strategies, each with advantages. Strategies used are specific to the businesses and even the products or services.

Bulk Purchasing	Direct Purchasing
Buying large quantities of a product at once, often to secure lower prices.	Buying products directly from manufacturers or suppliers, usually in smaller quantities and more frequently.
Typically offers significant discounts due to economies of scale.	Often higher cost per unit.
Requires sufficient storage space to accommodate large quantities.	Less storage space required.
Involves managing larger inventories, which can tie up capital.	Reduced inventory costs.
Slower responses to changes in demand and possibility of obsolete stock.	Offers more flexibility in purchasing decisions, allowing businesses to respond quickly to changes in demand.
Can maintain close relationships with suppliers, which can lead to better service and quality.	Can maintain close relationships with suppliers, which can lead to better service and quality.
Example: Pet shops might use bulk purchasing to stock up on essential items.	Example: Specialised exotic retailers might use direct purchasing to maintain fresh live feed and reduce storage needs.

Business term agreements

Business agreements are:

- Often known as **contracts** or **agreements**.

- They are **legally binding documents** that outline the rights, responsibilities and obligations of the parties involved in a business transaction.

- They are crucial for ensuring clarity, preventing disputes and providing legal protection.

- Details included in the agreements vary depending on the type of agreement.

For example, **sales agreements** and **service agreements**:

Remember!

Supply chain theories apply to any business which may be supplying stock or services. The animal management sector should always abide by business legislation as well as the legislation that prioritises animal welfare and ethical sourcing.

Sales Agreements:

- Parties Involved: Identifies the buyer and the seller.

- Products or Services: Describes what is being sold.

- Price and Payment Terms: Specifies the price and payment schedule.

- Delivery Terms: Outlines how and when the goods or services will be delivered.

- Warranties and guarantees: Details any guarantees provided by the seller.

Service Agreements:

- Scope of Services: Defines the services to be provided.

- Service Period: Specifies the duration of the agreement.

- Payment Terms: Outlines the payment structure.

- Performance Standards: Sets expectations for the quality of services.

- Termination Clause: Explains how and under what conditions the agreement can be terminated.

Important terms!

Supply chain: A network of businesses that are involved in getting a product or service to a consumer/customer.

Interdependency: The relying of two or more things on each other.

Ethics: Moral principles, policies and values.

Ethical sourcing: Sourcing of products, services or materials in a fair, environmentally friendly and socially responsible way.

Deadstock / Cadaver / Fallen stock: All different names for dead livestock.

Primary suppliers: Suppliers directly involved in the production of goods or delivery of services.

Business model: A plan that outlines how a business operates, makes money, and sustains itself in the market.

External influences: Things that happen outside of a business that it has no control over but may impact on the business.

Customer base: The group of customers who regularly buy or use the services of a business.

Procurement: The obtaining or purchasing of products or services for a business.

Business agreements: Legally binding documents that outline the rights, responsibilities and obligations of the parties involved in a business transaction.

Recap Questions

1. What is a supply chain?

2. What does CITES stand for and how does it impact a supply chain?

3. Why is efficiency important in a supply chain?

4. Why is ethical sourcing important in a supply chain?

5. What is supply chain assurance?

6. Give examples of different services within a supply chain.

7. What is an extended supply chain?

8. What are the consequences of failing to meet supply chain demands for a business?

9. What are the six key external factors affecting businesses?

10. What is a target market?

11. Who makes up a high value customer base?

12. What are indirect competitors?

13. Give the differences between direct and bulk purchasing.

14. Describe what service and sales agreements are.

4.2 Principles of consumables and stock management in the animal management sector

Principles

Stock means:

1. Any items kept by a business to use in production or to sell.

2. Can be raw materials, production elements or finished products, for example, wood and metal, mesh, cat carrier.

Consumables are items that are used or worn out and will require replacement. For example:

- feed
- wood shavings
- flea treatments.

Products in the animal management sector

Stock types

The **type of stock** depends on the business.

For businesses that supply or sell animals (pet shops, breeders), provide a service for animals (veterinary surgery, groomers) or have animal collections (zoos, aquaria) common stock items examples include:

- Feed for animals (specific to species) e.g. rabbit pellets.
- Bedding for animals (specific to species) e.g. wood shavings.
- Veterinary medicines and equipment / first aid equipment e.g. worming tablets, bandages.
- Animals themselves e.g. rats, rabbits and snakes.
- Husbandry equipment for the care and handling of the animals e.g. chicken coops, guinea pig hutch, cat carriers, dog muzzles, cleaning materials.
- Maintenance materials for accommodation and equipment e.g. nails, fencing wire, UV lightbulbs.

All stock can be divided into two main types:

- perishable stock
- non-perishable stock

Perishable stock

- Can only be **stored in the short term**, in certain conditions only and lose or expire value over time.
- Most obvious examples are feed, but it can also include anything with an expiry/use by date, such as medicines, first aid equipment and some bedding.
- Animals themselves are perishable stock.

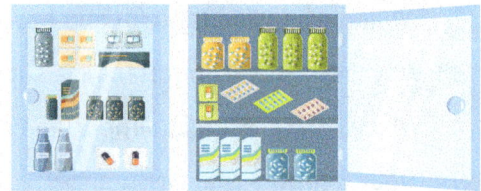

Medicines are an example of perishable goods

Non-perishable

- Can be **stored for a long period of time** without changing from original state or deteriorating.
- Examples include: accommodation, maintenance equipment, blankets, husbandry equipment, restraint equipment, grooming equipment.

Legislation, regulations and codes of practise impacting stock management

Legislation

- A **law** or set of laws that are passed by Parliament.
- Laws are also known as **Acts of Parliament**.
- Also refers to the process of creating laws.
- Includes bills which are pieces of legislation that are being considered for approval.
- Legislation can be enforced by the courts through penalties for those who breach laws/acts.
- For example – the Animal Welfare Act 2006 and its duty of care to provide the 5 needs.

Regulations

- **Reugulations** are **formal guidelines** that link to existing legislation.
- They **help to apply the principles of laws** and are supplementary to them.
- Breaching regulations is not necessarily enforced through the courts.
- For example –the Animal Welfare (Licensing of Activities Involving Animals) (England) Regulations 2018 which controls licensable activities such as selling animals as pets, boarding cats and dogs, breeding dogs and keeping and training animals for exhibition.

Codes of practice

Codes of practice:

- Are written guidelines from a professional body.
- Outline **standards and working practices**.
- Often link to legislations and regulations.
- A **breach of a code of practice is not necessarily an offence**. However, it can be used as evidence in a court case.

For example – Defra's Codes of practices for farm animal, dog, cats and horse welfare. These offer advice on owning the animals and link to the Animal Welfare Act 2006.

Impacts on stock management principles in the animal management sector

Animal Welfare Act 2006	Where live animals are the stock, it requires good practices of husbandry to maintain good stock levels within the business.
	Requires husbandry materials to be stocked at suitable levels to maintain animals or service for animals.
The Animal Welfare (Licensing of Activities Involving Animals) (England) Regulations 2018	Record keeping is required to show animal and supply purchases for licensing.
	Requires husbandry materials to be stocked at suitable levels to maintain animals or service for animals.
Defra's Codes of practice	Outlines recommendations for a suitable diet and housing conditions for animals.
	Requires husbandry materials to be stocked at suitable levels to maintain animals or service for animals.
Animal and Plant Health Agency (APHA)	APHA 'works to safeguard animal and plant health for the benefit of people, the environment and the economy.'
	It advises on standards of animal health, researches into animal diseases, monitors compliance with animal welfare regulations.
	Produces a number of guides for example notifiable diseases in animals and bringing your cat, dog or ferret into Great Britain.
	The guides outline requirements for husbandry which impacts on stock management.
Veterinary Medicines Regulations 2013	Sets out the controls on veterinary medicines and medicated feed.
	Describes storage requirements for particular medicines such as controlled drugs (see Animal Health).

Consequences of non-compliance with legislation

- Non-compliance means failing to meet or refusing to meet legislative or regulatory requirements or codes of practice.

- People or businesses can be **prosecuted** (a case brought against them in a court of law) for non-compliance which can result in fines, bans from keeping animals for a period of time or even a prison sentence.

- If the business requires **a license to operate then this can be taken away** because of non-compliance.

- A business with poor compliance will **lose reputation** either through reviews, word of mouth or due to publicity in media (including social media) which can result in a downturn in customers/clients.

- With animal management businesses and animal ownership any non-compliance can **impact on animal welfare**, leading to potential animal suffering.

Application of stock management for different business and purposes

Veterinary practices

Veterinary practices rely on effective stock management to ensure the smooth operation of clinics and patient care.

Types of stock include: medications, vaccines, medical supplies (e.g. syringes, bandages), food and surgical equipment.

Specific examples of stock management:

- **Inventory control**: monitoring usage of drugs and supplies to avoid shortages or overstocking.

- **Expiry date management**: ensuring medications and vaccines are used before expiration to prevent waste.

- **Emergency stock**: keeping critical medications or supplies (e.g. anti-venoms, emergency drugs) readily available for urgent cases.

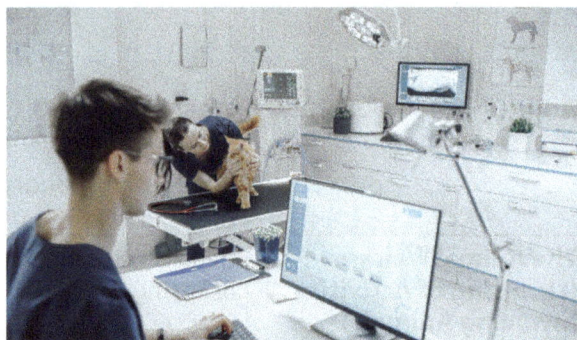

A veterinary clinic

Zoos

Zoos manage an inventory of live animals and items needed to care for these animals.

Types of stock include: animal feed (fresh produce, frozen meat, hay), veterinary supplies (medicines, vaccines, tools), enrichment items (toys, climbing structures) and maintenance supplies (habitat cleaning tools).

Specific examples of stock management:

- **Animal tracking**: managing animal stock through breeding programs, transfers or conservation initiatives.

- **Seasonal planning**: adjusting stock for breeding seasons or weather changes (e.g. increased food in winter for some species).

- **Data Systems**: using software to track animal health, diets and supply inventory.

Reputable breeders

Reputable breeders must manage stock effectively to ensure the health and well-being of breeding animals and offspring.

Types of stock include: animal food (specialised diets for breeding, nursing or young animals), medical supplies (vaccines, supplements, deworming treatments), bedding materials and cleaning supplies.

Specific examples of stock management:

- **Health tracking**: monitoring breeding animals and offspring for vaccinations and treatments.

- **Breeding cycles**: planning stock requirements around mating, pregnancy and weaning periods.

Pet store / pet shop

Pet stores manage stock to meet customer demand while ensuring animals and supplies are well-maintained.

Types of stock include: live animals (fish, birds, small mammals, reptiles), food and treats, pet accessories (toys, bedding, clothing, leashes), grooming products and health supplies (shampoos, medications).

Specific examples of stock management:

A pet shop

- **Demand forecasting**: adjusting inventory levels based on seasonal demand (e.g. holidays or summer pet adoptions).

- **Perishable goods management**: rotating stock like food to avoid waste.

- **Stock audits**: regular checks to monitor shelf life and prevent overstocking or shortages.

Boarding kennels/catteries

Boarding facilities manage stock to provide temporary care for pets, ensuring their comfort, health and safety.

Types of stock include: food (variety to suit dietary needs), cleaning supplies (disinfectants, bedding, litter), grooming items (shampoos, brushes) and medications for minor treatments or emergencies.

Specific examples of stock management:

- **Hygiene management**: maintaining cleaning supplies to meet hygiene standards.
- **Capacity planning**: managing inventory based on the number of pets in care.

Grooming

Grooming businesses manage stock to ensure pets receive quality grooming services, from basic cleaning to advanced care.

Types of stock include: shampoos, conditioners, skin treatments, grooming tools (clippers, scissors, brushes), consumables (towels, gloves, cleaning supplies) and accessories for pets (bows, coats, sprays).

Specific examples of stock management:

A dog grooming business

- **Inventory control**: monitoring high-turnover items like shampoos and disinfectants to avoid shortages.

- **Quality assurance**: sourcing products that are safe for pets (non-toxic, allergy-free).

• •

Methods of ordering and using stock

Stock rotation

Stock rotation ensures that stock items are sold or used in a correct order, with consideration for **expiry dates** (the date by which a product should be used by).

- The idea is to minimise waste, maintain quality and maximise efficiency.
- It is very important for perishable stock in particular, to help it stay fresh.

- The basic principle of stock rotation is first in first out (FIFO). For example, in a pet store FIFO ensures that the older stock is at the front of the shelf and the new stock goes into the back of the shelf.

Monitoring and maintaining stock levels to meet supply and demand

Managing inventory is a term used to describe monitoring and maintaining stock levels. It can also be known as a **stock take.**

It is important to manage inventory well to maintain the businesses success and competitiveness.

- Supply is how much of a product or services businesses are willing to make and sell at a certain price. If prices are high, then businesses will be willing to make and sell more as they can get more profit. If prices are low, then businesses may not want to make and sell as much as profits will be low.

- Demand is how much of a product or service people want and can buy at a certain price. If prices are high, fewer people can afford it, and demand may decrease. If prices are low, more people may want to buy, and demand may increase.

- **Ideally supply and demand should be in equilibrium with each other**. If supply is high and demand is low then prices may decrease. If demand is high and supply is low then prices may increase.

Stock needs to reflect supply and demand and be prepared for unexpected changes in supply and demand. There should be a minimum and maximum threshold for each item of stock.

- The **minimum threshold** is the least amount of stock that can be held before it impacts on the business.

- The **maximum threshold** is the most amount of stock that should be held before it impacts on the business. This could be dictated by storage as well as demand.

Dog coats are a seasonal product

Regular stock checks and maintaining records will help with predicting the flow of stock into and out of the business, whilst ensuring suitable cash flow in the business.

Seasonality of stock will impact on sales and hence stock ordering. Businesses need to look ahead to see what might be required for the coming seasons or for fluctuations in service demand. For example:

- Different feed may be available at different times of the year such as fruits and vegetables or cost differently in the off season (when not normally grown and importing may be required) at an animal collection.

- For a pet shop having products that match the seasons such as rain coats for dogs or suncream is important for sales.

- For a grooming parlour there may be greater demand during shedding seasons.

- For boarding kennels or catteries there may be a greater demand during the school holidays.

Maintaining records

Maintaining records is a vital part of stock management and safeguards animal welfare

- Monitoring stock levels should give an **indication of sales and help track trends** and patterns. It can lead to better planning and resource allocation, such as ensuring adequate supplies of food, bedding and medications.

- Records help **avoid under or overstocking** both of which can negatively impact on the business.

- Some **records are required by legislation** or for licenses to be valid, for example, controlled drugs in veterinary practices.

- **Records build customer trust and transparency**. For example, businesses selling or adopting out animals, providing accurate records (health certificates, breeding

Keeping records

history) builds customer trust. They assure buyers that the animals are well-cared for and meet welfare standards.

- Accurate records support the long-term success and sustainability of a business.

Implications of ineffective stock management

Ineffective stock management can impact a business in many ways, such as loss of income, loss of clients/customers, loss of reputation and sustainability. Here some examples:

Cost

Poor stock management can result in **increased expenses** and **reduced profitability**.

This can occur by overstocking or understocking which:

- Ties up money in excess supplies that may not be needed.

- Leads to additional storage costs (e.g. refrigeration for perishable items like animal food or medication).

- Can result in price reductions to clear stock.

- Creates operational delays, such as running out of essential medical supplies in veterinary practices or food for animals in zoos and boarding kennels.

Wastage

Ineffective stock management increases the risk of unnecessary waste, which harms finances, animal welfare and the environment.

Wastage can occur due reasons such as to expired stock, spoiled stock, over ordering and incorrect measurements.

It can lead to risks of running out of supplies such as food, medicine or enrichment materials can harm animal health and well-being, especially in zoos, breeders and boarding kennels.

Compliance with relevant legislation and codes of practice

Stock management is often **governed by legislation, regulations and codes of practice**, especially when animals, medications or perishable goods are involved.

Examples of issues include

- Running out of stock (e.g. medications or enrichment tools) can result in a breach of welfare laws, as animals may not receive adequate care.

- Inadequate food supplies or poor-quality stock can violate codes of practice regarding animal welfare.

- Failure to dispose of expired drugs or spoiled food properly can lead to legal penalties or fines.

- Poor stock management may affect the ability to trace sources of products, such as food, medicines or animals. This can breach regulatory requirements for businesses like breeders or pet stores.

- Non-compliance with hygiene standards for stock storage (e.g. cleaning products or food) can pose health risks to animals, staff and customers.

Non-compliance can lead to fines, damage to reputation, loss of customer trust and a negative financial impact on the business.

· ·

Methods of storing products

Storage requirements

- It is important to ensure that **stock is stored correctly** as per the manufacturer's instructions.

- This is particularly important for **perishable goods**.

- **Space availability** may dictate quantity of stock stored especially larger items.

- Storage systems can vary depending on items; shelves, racking, crates, fridges or freezers. They can even be automated.

- Stock should be **clearly labelled** and **easily accessible**. Lifting equipment should be provided if needed e.g. forklifts and trolleys and safe lifting training given to staff.

- Considerations include; temperature, sunlight, lighting, humidity, orientation, stacking, weight, security, hazardous materials, contamination.

- **Storage costs and sustainability** are important considerations when planning stock storage. Some sector-specific examples include:

  » Some veterinary medicines must be stored in a fridge with expiry dates. Controlled drugs must be stored in accordance with the Misuse of Drugs Regulations 2001.

  » Raw dog food should be stored in a fridge or freezer as appropriate, and contamination of other stock avoided.

  » Sheep nuts should be stored in vermin proof bins on a farm to avoid loss of feed and contamination.

  » Dog food bags must be stored carefully in larger sizes in a pet shop, so they are not a hazard for staff to lift them onto shelving or for customers taking them off the shelves.

Different ways to store products

Animal welfare

- **Live animal stock** is subject to **animal welfare legislation** whether it is being sold for food, breeding or for collections.
- The **Animal Welfare Act 2006** states that the responsible person has a duty of care to take

reasonable steps to provide the 'five needs' for the animals in their care.

- Remember live vertebrate prey cannot be fed to animals in the UK.

Dealing with deliveries

Deliveries of stock can vary including depending on type of stock ordered, its transport requirements and size.

- Stock should be checked against what was ordered as well as its condition and quality.
- Stock should be dealt with promptly and stored in accordance with each item's requirements.
- Deliveries should be scheduled during business hours to ensure that stock can be received.
- Health and safety must be observed when unloading and putting away deliveries.
- Live animals must be carefully dealt with at delivery to minimise stress and reduce poor welfare.

- Isolation and quarantine requirements should be considered, to protect from contagious/infectious disease.
- Animal transport certificates and movement forms should be completed if required.
- Housing should already be prepared for the animals so that there is no waiting in inappropriate conditions.
- Consideration should be given to travel times and feed and water requirements.
- Loading and unloading facilities should be used that are designed and constructed to avoid injury and suffering.

Dealing with deliveries

Recap Questions

1. What maintenance equipment might be required in a pet shop?

2. What is the difference between perishable and nonperishable stock?

3. What is the difference between legislation, regulations and codes of practice?

4. Give examples of legislation that impacts on animal management business stock.

5. Why might zoos use data management systems for stock control?

6. Why is stock rotation important?

7. Explain minimum and maximum thresholds of stock.

8. How does seasonality impact on stock ordering?

9. What are the consequences of poor stock management?

10. Describe the considerations required with storage of stock.

11. What are the health and safety requirements of dealing with deliveries of stock?

Practice Questions

1. Describe, with examples, what makes up an animal management supply chain. (6 marks)

2. Explain how legislation can impact on animal management supply chains with named examples of laws. (5 marks)

3. Describe the interdependency of suppliers, distributors and customers within a supply chain for a pet dog's drying coat. (5 marks)

4. List and explain two benefits of ethical sourcing of stock for an animal bedding wholesaler. (4 marks)

5. Define fallen stock services and describe the services available for it. (3 marks)

6. State which organisation provides accreditation for pet crematoriums. (1 mark)

7. List three supply chain sequences and describe their operation. (6 marks)

8. Compare and contrast primary, secondary and tertiary suppliers giving an animal management business example. (6 marks)

9. Explain how staff wellbeing can be impacted by a business failing to meet supply chain demands. (5 marks)

10. Explain how socio-cultural factors can impact on a pet shop business. (5 marks)

11. Define market segments and give an animal management business example. (3 marks)

12. Give four examples of indirect procurement for an animal collection. (4 marks)

13. List two stock items that would be required in a veterinary practice and describe their use. (4 marks)

14. List three perishable stock items for an equine yard and the explain impact of incorrectly storing these items on the business. (6 marks)

15. Explain the differences in the consequences of not meeting legal requirements, regulations and codes of practice. (5 marks)

16. Explain how the Veterinary Medicines Regulations 2013 impacts on stock control at a veterinary practice. (4 marks)

17. Describe the impact that Defra and APHA have on stock management for a guinea pig breeder. (5 marks)

18. Describe how efficient stock rotation can impact on reducing wastage in a pet shop. (4 marks)

19. Explain why maintaining stock records is important for a reputable pedigree domestic cat breeder. (5 marks)

5.1 Principles of learning theories

Understanding animal behaviour and the ways in which they learn is key to good animal husbandry and welfare. This is true whether animals are domestic, in a collection, or being trained for a specific purpose.

The **Animal Welfare Act 2006** and **Animal Welfare (Licensing of Activities Involving Animals) (England) Regulations 2018** must always be considered when animals are being housed and trained with respect to their behaviour.

Animal sentience is the principle of animals having the capacity to experience positive and negative emotions and is a term applied to vertebrates, cephalopods (octopus, squid etc.) and decapod crustaceans (lobsters, crabs etc).

The Animal Welfare (Sentience) Act 2022 allows vertebrate animals, decapod crustaceans and cephalopod molluscs to be recognised as sentient. It requires the consideration of animal sentience when laws are being proposed or changed, and the establishment of an **Animal Sentience Committee** to help guide government policy in relation to animal welfare.

Octopus are considered to be sentient

Key learning theories

Learning takes place during an individual animal's lifetime. By understanding how animals learn in the wild, their behaviour in collections, domestic settings and in work can be shaped to improve welfare and train for higher learning.

Types of learning

There are key **higher learning theories** to be understood before applying learning theories to training.

Social learning

- Animals that live in social groups can **learn behaviours from other members of the group**, whether this be by interactions or observation.

- Examples cover behaviours such as food selection, foraging, predator avoidance, route learning, song learning and motor skills.

- It can allow the spread of new behaviours within a group. **Cultural behaviours** which persist over generations can be learned in this way, such as potato washing by Japanese macaques.

- **Trial and error** can also contribute to social learning.

Social learning

Observational learning

- An animal can learn by **observing the behaviour of others**.

- For example, young animals often learn foraging and hunting behaviours through **observing and copying the behaviour of older animals**.

- As with social learning, **trial and error** can also contribute to observational learning, where the animal will make repeated attempts at the task in varied ways until achieving success.

Latent learning

Latent learning is also known as **incidental learning**, and it can occur in the absence of other animals.

- The animal acquires knowledge by observing something, but may only use that knowledge later on. There is no immediate reward or action in latent learning.

- An example is an animal exploring its surroundings and learning the location of its home area. Later, this could prove important for finding food or escaping a predator.

Insight learning

Insight learning requires the animal to solve problems by viewing a new situation as a whole and mentally working out the solution rather than using trial and error learning to achieve the outcome. The animal will suddenly 'see' the solution to a problem.

- This type of learning has been observed in chimpanzees, rats, and rooks.

> ### Important terms!
>
> **Learning** - A permanent change in an animal's reaction to a stimulus or situation because of experience, observation or training.
>
> **Sentience** - The capacity to experience positive and negative emotions.
>
> **Higher learning** - The process of acquiring knowledge through learning such as social, observational, latent and insight learning.

Insight learning

Cognition

Cognition refers to the mental processes in the mind that lead to knowledge and understanding. These mental processes include the use of memory and information from the senses.

Real world example

- Brown tufted capuchins in Brazil can take eight years in the wild to learn how to remove palm nuts from their casings.

- The young capuchins observe adults performing the behaviours and then trial and error their own behaviours to perfect the process.

- They learn to take off the fibrous husk from the palm nut, leave it for one week to bake in the sun, tap it to see if it is ready to be eaten, find a suitable flat rock to act as an anvil to lay the nut on, and finally choose a suitable different harder rock to use as a hammer to break the nut open for its rich oily kernel.

- This process requires intelligence, planning and dexterity.

A capuchin using a rock to break a nut open

Importance of captive animals' learning natural behaviours

The Animal Welfare Act 2006 requires a duty of care to provide the five needs for captive animals. One of these is the need to be able to exhibit normal behaviour patterns.

- **Normal behaviour patterns** will include those seen in the natural environment.

- For captive animals who are being released into the wild, either through rehabilitation or reintroduction, **learning natural behaviours is essential** for survival.

Releasing a seal cub

Innate behaviour

Whilst animals must learn some behaviours, there are certain other behaviours they are born with and do not need to learn.

- **Innate behaviours** are those which are not learnt and are the result of genetics. They can also be called **instinctive behaviours**.

- Innate behaviours are a response to a stimulus and usually show little variation between individuals.

- For example, adult herring gulls will regurgitate food for their chicks when the young peck the red spot on parent's beaks. Greylag geese will roll an egg back into the nest if it slides out.

Adult herring gull

Socialisation

Socialisation with other animals of the same and different species is crucial at certain stages in life, so that the animal is able to react suitably as an adult in social interactions. Examples include mating, predator / prey interactions and conflicts.

- With domestic species such as cats and dogs, there is a socialisation period, where it is important for them to have a wide range of experiences with different situations, other animals, and people, so they learn to react suitably as they mature.

- For kittens, this period is from birth to 7/8 weeks of age.

- For puppies it is approximately 3-17 weeks depending on breed.

Socialisation

Nature/nurture

An animal's behaviour is usually a combination of **innate** and **learned behaviour**.

- **Evolution shapes innate (nature) behaviour** via genetics and experiences within a lifetime

- **Nurture shapes the development** of learned behaviour.

The **environment around an animal has an impact on captive animal's behaviour**. This is true of wild animals in a collection or domesticated animals.

It is important for animal welfare to understand that nature and nurture both influence animal behavioural development.

Imprinting

Imprinting is a bonding process whereby young animals learn the characteristics of other individuals early in life and preferably associate with them thereafter.

- It takes place in a sensitive period, often shortly after birth.

- There are two types of imprinting.

  » **Filial**. Young animals learn characteristics of the mother and stay close to her so that they do not lose her and her care for them, a survival strategy. Occurs in many species of birds and some mammals.

Filial imprinting

- **Sexual**. Young males imprint on the appearance of their mother to help with mate choice later in life. Ensures they select the correct mate in terms of species and hence survival of genes via reproduction. Occurs in some bird species, for example finches, turkeys and gulls.

Hand-rearing by humans can cause an animal to imprint on them instead. This can cause problems with mating, or when reintroducing or releasing animals back into the wild after rescue. Care needs to be taken to disguise the human carer as much as possible or to minimise human contact.

Sexual imprinting

Other factors influencing behaviour

Innate behaviours are shaped by genetics and therefore evolution. Natural selection will ensure behaviours that best fit the existing environment will be passed on to future generations through 'survival of the fittest'.

Hand rearing disguise

A variety of other factors may influence the behaviour of an animal, such as life stage, health status, diet and emotional state. These factors are all part of the nature / nurture aspect of behaviour and must all be considered when studying behaviour, caring for animals and training animals.

For example:

- Elderly animals may become more withdrawn or aggressive due to pain from conditions such as arthritis.

- Young animals and domesticated animals tend to be more playful than their adult or wild counterparts.

Young goats playing

> ### Important terms!
>
> Innate behaviour: Behaviour shown that is from the genes inherited by an animal from its parents.
>
> Socialisation periods: Crucial time in a young animal's life where its experiences or lack of them shape its future behaviour.
>
> Nature/Nurture: The combining forces of genetics (nature) and the environment (nurture) on an animal.
>
> Imprinting: The bonding of an animal to its parent for survival purposes.

> ### Remember!
> The behaviour of an animal is more than just innate and learned behaviours, it is influenced by many factors and can be individual to that animal. It is important to understand the behaviour of a species **and** an individual before training any animal.

Real world example

Konrad Lorenz studied the imprinting mechanisms of geese in the 1930s. A clutch of eggs was divided into two halves close to hatching. One half remained with the mother and the other half were placed in an incubator. Lorenz made sure he was the first moving object that the incubator goslings saw once they hatched. He imitated the mother's call and found that the incubator goslings would follow him, and the nest goslings would follow their mother. He even combined up the two groups of goslings and found they separated to follow their respective 'mothers' – him and the female goose. Lorenz believed the imprinting was innate and genetic and could not be reversed once it had occurred.

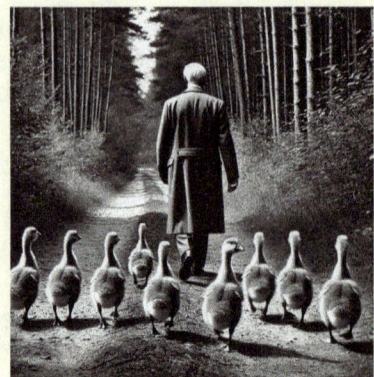

Voluntary behaviour, involuntary behaviour and training theories

- **Voluntary behaviours** are those that an animal has control over performing or not.

- **Involuntary behaviours** are those that the animal has no control over performing or not, they occur automatically. They are essential for survival and maintaining the body's internal balance (**homeostasis**).

- Training can make use of these behaviours in wild, zoo and domestic situations.

Non-associative (stimulus response) learning

This is a basic form of learning where an animal learns to **react to a stimulus** with a **persistent change**. The reaction can be influenced by genetics and early experiences. There are two main types of non-associative learning:

- Habituation: this is where an individual's reactions to a stimulus decrease or stop after repeated or constant exposures to the stimulus. It helps animals filter out the non-threatening (neutral) stimuli from its environment. Neutral stimuli are those that have no positive or negative impact on the animal and hence survival and responding would be a waste of time or energy.

  » **Wild Animal Example** – birds soon learn to ignore a scarecrow which prevented them from landing in a field when first placed there.

  » **Domestic Animal Example** – a dog stops barking at a seagull approaching on the beach on its daily walk.

  » **Zoo Animal Example** – zoo animals can habituate to enrichment and lose interest in it.

  » **Use for training** – animals trained to 'learn' to ignore any stimulus in their environment that has no consequence for them, e.g. visitors to a zoo

- Sensitisation: This is where an individual's response to a stimulus increases from repeated presentation. It is the opposite response to habituation. It is usually in response to significant stimulus, such as presentation of food or a predator. Sensitisation does also play a role in classical conditioning (see later).

  » **Wild Animal Example** – Harbour seals reacting to threatening orca underwater calls rather than non-threatening orca calls

  » **Domestic Animal Example** – dogs becoming agitated with the sound of fireworks

  » **Zoo Animal Example** – primates becoming withdrawn due to high visitor numbers

  » **Application** of this type of learning for animals – there are obvious benefits to

A dog afraid of fireworks

this type of learning in domestic, wild and zoo situations. Habituating to non-threatening stimuli, such as visitors for domestic and zoo animals, is key for animal welfare. Sensitisation to stimuli can cause poor welfare that will need to be remedied and yet at the same time can help in training animals such as in zoo visitor displays.

- **Desensitisation:** This is a behaviour modification technique that is used to get animals used to certain stimuli, such as loud noises, unusual noises, or other animals. A trigger stimulus has been identified that the animal needs to desensitise to, and it starts with exposing the animal to a low level of the stimulus so that response is minimal. Over time, the stimulus level is increased until there is no response at full intensity. Positive reinforcement can be used alongside the increasing intensity to aid the process.

  » **Example** – desensitising to a vacuum cleaner noise for a dog would involve initially muffling the noise by being enclosed in another room and gradually increasing the noise level until the dog can be in the room with the vacuum cleaner without reaction.

Classical conditioning

Classical conditioning is a form of associative learning. It is also known as **Pavlovian conditioning** after Ivan Pavlov, who first studied this type of conditioning in late 1800s. It can also be referred to as **respondent conditioning**.

In this form of learning, an animal **forms an association** between a significant stimulus, the unconditioned stimulus (UCS) and a neutral stimulus (NS).

- The reaction to the UCS is called the **unconditioned response (UCR)**
- The reaction to the NS alone is called the **conditioned response (CR)**.

By repeatedly exposing the animal to the paired UCS and NS, the animal can go on to produce the UCR to just the CS alone.

Pavlov demonstrated this by studying the production of saliva in dogs in response to food.

The sight, smell and taste of food cause the production of saliva in dogs. Pavlov introduced the ringing of a bell before food was presented. Initially there was little reaction from the dogs to the bell. However, over repeated bell and food pairings, they began to salivate on the ring of the bell, before food was presented. In other words, they produced a **conditioned response** to the bell, which was a previously neutral stimulus.

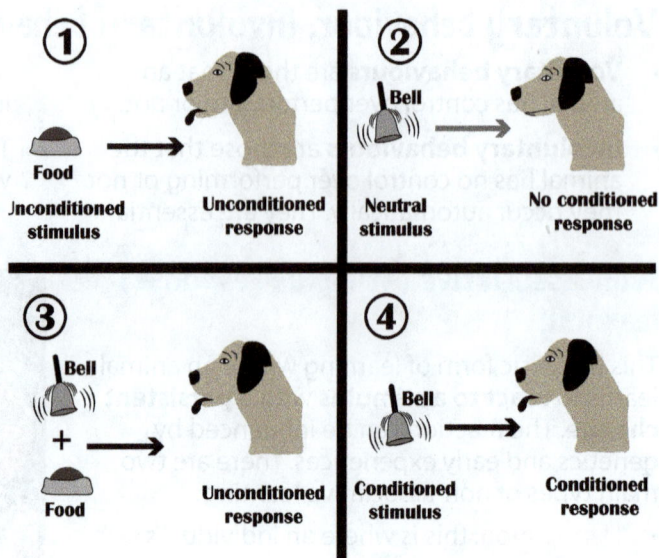

Pavlov's experiment

Examples in real life:

- Dog getting excited when their owner picks up the lead, as they have learnt to associate the lead with walks, a positive outcome.
- Cats getting anxious and run away when they see a cat carrier, because they have learnt to associate the carrier with going to the vets and possible negative outcomes.
- Clicker training is where a neutral stimulus (NS) is paired with a positive stimulus (UCS) such as a treat. Commonly used in domestic animal training and in the wider zoo/captive animal environment.

Operant conditioning

Operant conditioning is a form of associative learning. It is also known as **instrumental conditioning** and is based on **trial-and-error learning**. Edward Thorndike and Burrhus Skinner both carried out studies on operant conditioning. Skinner is more widely known for the theory however he based his work on Thorndike's 1898 Law of Effect.

Thorndike's Law of Effect

The animal learns to associate an outcome (such as a reward) with performing a behaviour rather than a particular stimulus. Animals can also learn not to do things by operant conditioning however this is from associating a negative outcome (such as a punishment) with performing a behaviour.

This process is also known as **shaping**.

Thorndike's studies looked at domestic cats and their ability to use a lever to escape from a puzzle box. He worked with hungry cats and food outside of the puzzle box. The cats worked with trial-and-error attempts to exit the box and would eventually press the lever to open up the box and reach the food. In repeats of the trial, the cats would become progressively quicker at using the lever. Thorndike's Law of Effect suggests that any behaviour followed by a positive consequence is likely to be repeated and any followed by a negative consequence is likely to cease.

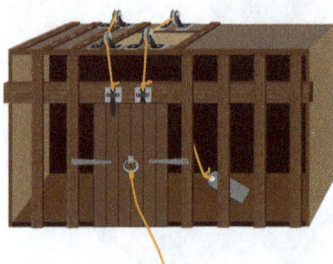

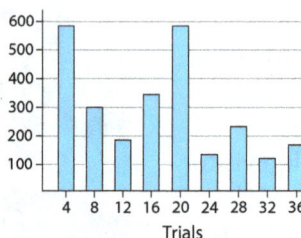

Thorndike's puzzle box experiments and Law of Effect

Skinner Box

Skinner (1938) followed on from Thorndike's work studying this topic using a similar box, now called a **Skinner box**. (It is also known as an **operant conditioning chamber**).

The animal can be rewarded with food (positive consequence) for performing certain behaviours e.g. lever pressing in rats or key pecking in pigeons.

Alternatively, the animal can press the lever/peck the key to stop punishment (positive consequence) in the form of electric shocks. As with the cats, initially there is a trial-and-error process before the animal learns to perform the behaviour required.

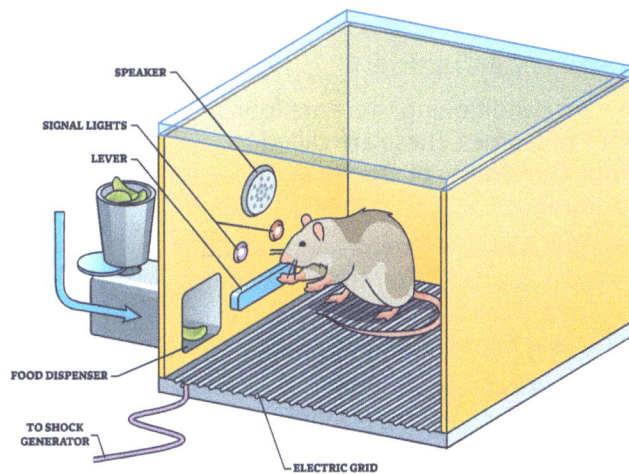

SPEAKER
SIGNAL LIGHTS
LEVER
FOOD DISPENSER
TO SHOCK GENERATOR
ELECTRIC GRID

A Skinner Box

Real-life examples

- Dog show training – each of a dog's movements, postures and responses are trained by the handler using instrumental training with food rewards.

- Zoo stations – animals in zoos are trained to go to a 'station' with food rewards. Initially, they are lured there by food and when they reach the station, they get a food reward. The animal will then learn to go to the station. This is useful for getting an animal to a certain spot in its enclosure so a variety of reasons such as visual health checks, freeing up a section of the enclosure or for visitor displays and talks.

These are both examples of direct stimuli that are external to the animal.

Internal/indirect stimuli

Internal/indirect stimuli do impact on the learning process and can limit the extent of operant conditioning.

- For example, it was attempted to train pigs to pick up a wooden coin and drop it in a container. They could be trained to pick up the coin but reverted to dropping it and rooting it, a more instinctive/innate behaviour.

- Similarly, a male three-spined stickleback could be trained to swim through hoops to reach a female but not to bite a glass rod to gain access to the female as they kept directing courtship behaviours to the glass rod.

In addition, internal/indirect stimuli can influence operant conditioning in terms of motivating the animal. For example, Thorndike

Dog show training

A lemur at a zoo station

Pigs rooting

used hungry cats in his puzzle box research, as he found they were more motivated to 'escape' if they were hungry and food was outside of the box. Equally, with Skinner boxes, rats are more motivated to press the lever to 'escape' when subjected to the pain of the electric shocks.

Applicant of quadrants of operant conditioning

In operant conditioning there are four possible outcomes. These are called the **Four Quadrants of Operant Conditioning**:

- **Positive Reinforcement** – something pleasant happens after the behaviour to increase the likelihood of the behaviour being repeated

- **Negative Reinforcement** – something unpleasant is removed after the behaviour happens to increase the likelihood of the behaviour being repeated

- **Positive Punishment** – something unpleasant happens after the behaviour to decrease the likelihood of the behaviour being repeated

- **Negative Punishment** – something pleasant is removed after the behaviour happens to decrease the likelihood of the behaviour being repeated

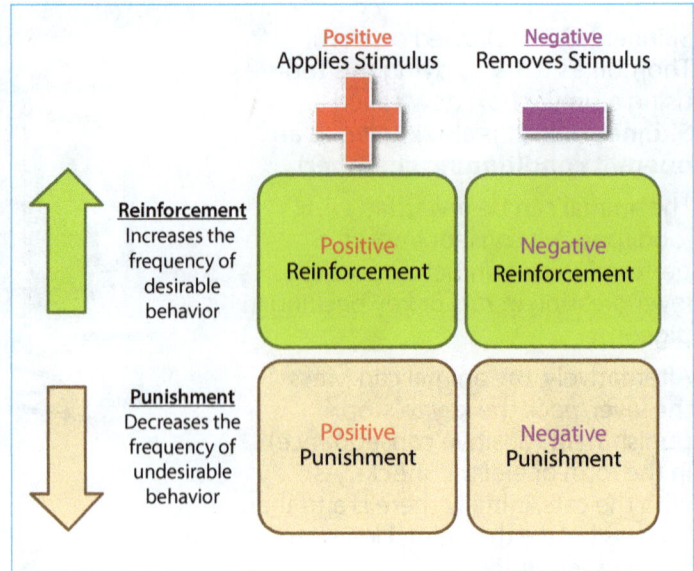

	Positive Applies Stimulus	Negative Removes Stimulus
Reinforcement Increases the frequency of desirable behavior	Positive Reinforcement	Negative Reinforcement
Punishment Decreases the frequency of undesirable behavior	Positive Punishment	Negative Punishment

Operant conditioning

Examples of the four quadrants in action

Positive reinforcement

Dog sits and receives a food treat for sitting, to increase the sitting behaviour

Negative reinforcement

Dog returns to owner's side when lead walking, as pressure from restraint lessons on muzzle, to increase lead-walking by owner's side

Positive punishment

Spraying a dog with water when it barks at people passing the house, to decrease the barking behaviour

Negative punishment

Owner turns away from dog when it jumps up, removing attention in order to decrease jumping up behaviour

5 Learning theories

It is thought that all four quadrants have their role to play in animal training. However, it does depend on the animal, their environment and where they are in their learning journey.

- **Positive reinforcement** is a good focus for training as it follows the LIMA approach (see below).

- **Continuous reinforcement** is more effective in establishing a particular response – reinforcing the behaviour every time. For example, a reward every time a lever is pressed.

- **Partial reinforcement** is more effective to maintain a response – reinforcing the behaviour every so many times or after a set period of time. For example, every third press of the lever or every 3 minutes of lever pressing. It avoids extinction of the response, which can occur if the response is not rewarded. For example, if there is no reward ongoing for pressing the lever then the response will stop.

> ## Remember!
> - It should be remembered that **positive** means **adding something** and **negative** means **taking something away** rather than good or bad.
> - **Reinforcement** is about trying to increase the frequency of a behaviour occurring.
> - **Punishment** is about decreasing the frequency of the behaviour occurring.

- **Aversive stimuli,** such as removing something unpleasant (negative reinforcement) or having something unpleasant occur (positive punishment) can cause emotional issues for the animal and concerns for welfare. They can also lead the animal to associate the aversive stimuli with training and the trainer, which can have negative consequences.

Least Intrusive, Minimally Aversive (LIMA) Approach

- **LIMA** stands for **least intrusive, minimally aversive** and is a type of animal training.

- It is based on the idea of using humane and effective methods to achieve the required behaviour outcome.

- It is based on positive reinforcement and not on the use of punishment, aversive tools and methods or physical force to minimise stress or harm.

- It aims to preserve the animal's dignity and comfort while encouraging positive behaviours, respecting the animal's welfare.

- It considers the animal's history and asks what outcome is required.

- It follows the humane hierarchy of behaviour change.

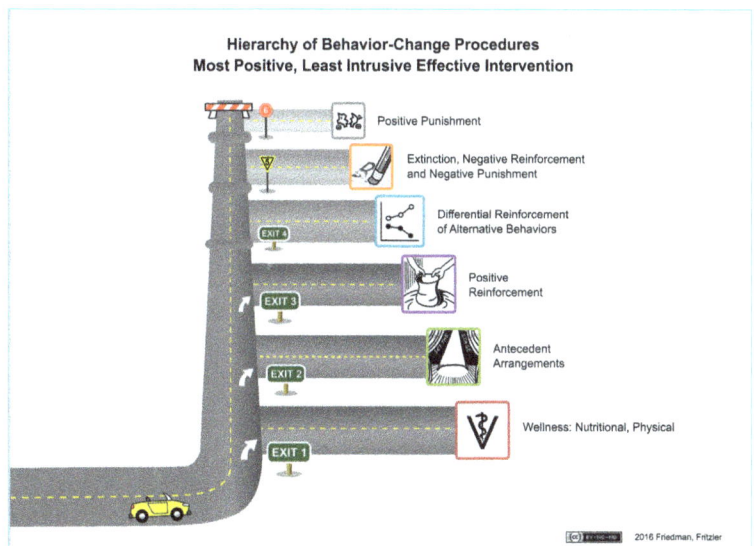

Figure 5.1.30 Humane Hierarchy of Behaviour Change

Steps for Implementation:

Start with the least intrusive technique: Use simple, natural interventions first (e.g. rewarding desired behaviours, redirecting attention).

Escalate only when necessary: If the mild techniques don't work, consider slightly stronger methods, but only as a last resort.

Use positive reinforcement: Reward desirable behaviours (e.g. treats, praise, play) instead of focusing on punishing undesired behaviours.

Minimise aversive tools or techniques: Avoid using punishment or physical corrections unless absolutely necessary.

Appropriate techniques

- Training programmes should follow a **cycle of planning, assessing and reviewing**.

- The **natural history of the species** should be considered.

- They should be **tailored to the individual animal**, its history and needs and should follow LIMA.

The following cycle is ideal for a training programme (SPIDER):

The **learning environment is also important**, and it may be that the programme needs to work over different environments before it can be carried out in the goal environment successfully.

- For example, recall training for dogs needs to start in the home environment and then move to a garden, then to a small outside space etc. Successful achievement has to be built up in increments before a dog can

Setting Goals

Planning

Implementing

Documenting

Evaluating

Re-adjusting

be considered to have solid recall in most environments.

Common training terminology

Term	Definition
Station/Target	Place or item that the animal is to reach or touch. For example, a perch or pointer.
Cue	A stimulus that elicits a behaviour. Cues may be verbal, physical (e.g. a hand signal), or environmental (e.g. a curb may become a cue to sit if the animal is always cued to sit before crossing a road).
Reinforcer	A stimulus which when presented to an animal, increases the likelihood of a behaviour occurring.
Clicker	A noisemaker. Animal trainers make use of the clicker as an event marker to mark a desired response. The sound of the clicker is an excellent marker because it is unique, quick and consistent.
Luring	A hands-off method of guiding the animal through a behaviour. For example, a food lure can be used to guide an animal from a sit into a down.
Baiting	The act of encouraging an animal to perform a desired behaviour through the static placement of a reinforcer. For example, placing a food treat on a target or station.
Moulding	Hands-on training that involves physical manipulation of the animal to perform a desired behaviour then reinforcing it. For example, pushing a dog's back end down into the sit position.
Shaping	Reinforcing successive appropriate approximations of a desired behaviour. Successive approximations are any behaviours that get the individual closer to the desired behaviour.
Extinction	The decreasing of behaviour through non-reinforcement or 'ignoring' the behaviour.

Training a horse

A clicker

Real-world examples of training animals

Working animals: Search dogs

These dogs are initially trained to play and search for a tennis ball that gradually has a scent added to it. This exploits their natural behaviour. It is self-rewarding in that they find their 'toy'.

Over time, this search behaviour is trained so that the dog will indicate when they have found the desired scent and in return receive their 'toy' to play with. It follows positive reinforcement.

Search dogs are used in the UK by the Police, Border Agency, Airport Security (see photo, top right) and Mountain Rescue (see photo, right). Commonly used dog breeds are English Springer Spaniels, Cocker Spaniels, Labradors, Border Collies and Beagles.

Scents that they can be trained to detect include:

- Class A Drugs: Cocaine, Heroin, Amphetamines
- Tobacco: Cigarettes, Hand-Rolled Tobacco
- Cash: Sterling, Euros, Dollars
- Product of Animal Origin (POAO): Meat, Fish, Honey, Dairy Products
- Firearms
- People: alive, cadavers, underwater
- Disease: COVID, epilepsy, cancer

Search dogs are used widely around the world in disaster zones such as after avalanches, earthquakes and for detecting landmines.

Other animals are trained too – for example, African giant pouched rats are trained to detect landmines (see photo, right).

Wild animals

Wild animals can be trained to improve with their day-to-day survival. This includes those that may have been released into the wild from captive breeding programmes.

Wild **chimpanzees** in Sierra Leone were trained to 'scream' as a whole troop in response to the sight of humans, to help protect them from poaching. (See photo, right)

- Sentinel chimpanzees would naturally scream when poachers (humans) approached. However the local rangers could not hear them from their station as it was too far away.
- Pipes that dispensed fruit and insects were set up for the chimpanzees, and when a person approached a remote-control button was used to release the food. The chimpanzees learnt that if they screamed when a person approached, they got food.
- The noise of the whole troop screaming was clearly heard at the ranger station, and there was a significant decrease in poaching as a result.

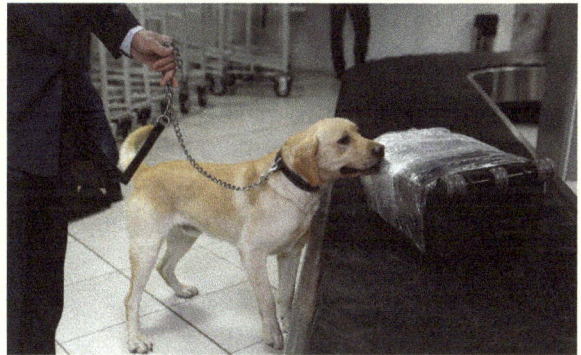

Wild **polar bears** have been trained to keep away from human settlements in the Arctic.

- The traditional method of using guards and firecracker shells to scare off polar bears coming into towns was ineffective long term, with some towns experiencing over 300 encounters annually.
- The solution was in three parts to create a lasting change in behaviour.
- Firstly, to time the scare tactic when the bears

interacted with human-related objects (e.g. trash cans and fences) so that the bears learnt to avoid those areas (positive punishment). This was more successful than trying to scare them when they were only approaching town. The goal was to make human areas undesirable without harming the bears.

- The second part was to make accessing human food harder with better secured trash cans, etc.

- The third part was to offer easier food sources in the wild, away from town – positive reinforcement for visiting these areas. Overall, these approaches reduced bear encounters for some towns to fewer than 10 per year.

Zoo animals

Training **zoo animals** can help with their day-to-day husbandry, involvement in keeper talks and displays. It can also help those in captive breeding programmes who are going to be released into the wild

- Baiting or luring to stations can be useful for public displays, as well as for health maintenance such as checking weight and parts of the body (such as teeth).

Clickers and targets are used to position animals so that checks can be made in the least intrusive manner possible. For example:

- Polar bears can be trained to targets so they can open their mouth for teeth checks.

- Big cats can be trained to a station in their enclosure, so that their tail can be looped out of the enclosure for blood to be taken, reducing the need for anaesthetics.

- Ring tailed lemurs being trained to separate stations for feeding so that medication can be given to one of the group as required.

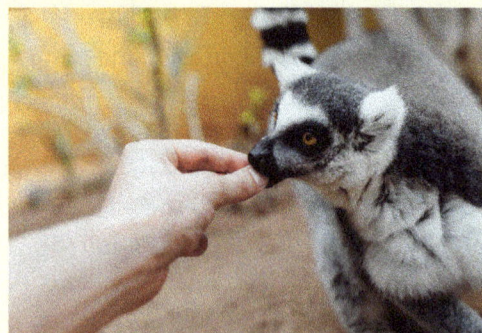

Domestic animals

Various animals can be trained to walk on a lead and harness – not just dogs. These include cats, rabbits, ferrets, goats, cattle, alpaca, hamsters and, of course, horses. This does all require positive reinforcement and gradual steps to accept the harness and to walk with the lead.

More complicated training programmes can help with dressage in horses, rabbit show jumping and dog displays.

Cats are not normally trained as much as dogs, though training them to go in a cat carrier is probably one of the most useful things an owner can do. Clicker training, or target/station training, could be used for this.

Rescue/stray animals

Many dogs and cats that end up in rescue centres have been handed in by their owners or abandoned/lost as stray or seized, having been abused.

These animals may need to learn to associate humans and handling with positive experiences, and this will take time.

For example, research has shown that positive handling and interactions with cats, during a stay in a rescue and re-homing centre, can increase positive reactions to strangers, and therefore increase their chance of being re-homed.

Clicker training and positive reinforcement of desired behaviours can help with training both dogs and cats towards the goal of re-homing. It can also help with decreasing undesirable behaviour, another benefit to re-homing.

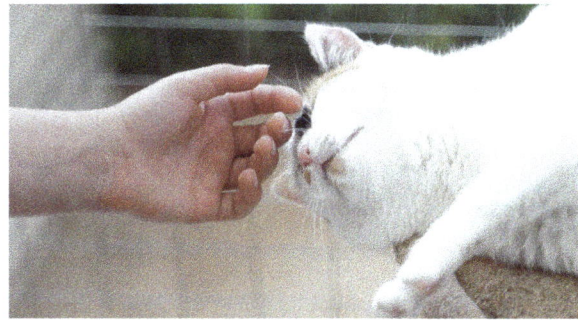

Important terms!

Habituation: Reactions to a stimulus decrease or stop after repeated or constant exposure.

Sensitisation: Reactions to a stimulus increase due to repeated presentation.

Classical conditioning: Associative learning where an animal learns to produce an unconditioned response to a conditioned stimulus.

Operant conditioning: Associative learning where an animal learns to associate an outcome with performing a behaviour.

Remember!

A training programme should be tailored to an individual animal, its history and needs. Following the SPIDER cycle and abiding by a LIMA approach is the best way to consider an animal's welfare needs during training. Training programmes may take time and perseverance to reach their goals.

Recap Questions

1. Define animal sentience.

2. Which legislation must be considered when training animals?

3. Define social learning.

4. What is meant by cognition?

5. What is innate behaviour?

6. When is a dog's socialisation period?

7. What is meant by nature / nurture?

8. How does imprinting affect hand rearing?

9. What factors can also influence behaviour?

10. What are habituation and sensitisation?

11. How does classical conditioning work?

12. What are the four quadrants of operant conditioning?

13. What does LIMA mean for animal training?

14. The SPIDER cycle of training contains which steps?

Practice Questions

1. Explain the difference between latent and insight learning with named examples. (4 marks)

2. Describe observational learning with a named example. (3 marks)

3. Explain the difference between innate and learned behaviour. (4 marks)

4. Describe the kitten's socialisation period and its importance for adult life. (4 marks)

5. Explain why an animal's behaviour is a combination of nature and nurture. (4 marks)

6. Name the two types of imprinting and explain their importance. (4 marks)

7. Describe what factors might influence an adult dog's behaviour. (5 marks)

8. Define habituation and give an example in a pet animal. (3 marks)

9. Explain how a named animal can be trained to desensitise to loud noises that they find distressing. (3 marks)

10. Describe Pavlov's studies with dogs with explanations of the types of stimuli and responses involved. (6 marks)

11. Explain how zookeepers use operant conditioning to help with visual health checks and keeper talks, giving two named examples. (5 marks)

12. Name the four quadrants of operant conditioning. (4 marks)

13. Explain how positive reinforcement can be used to train animals with two named examples; one in a working animal and one in a domestic animal. (4 marks)

6.1 The structure and function of the digestive system in relation to animal physiology

Throughout the animal kingdom, the digestive systems of different species have evolved to enable animals to utilise the food sources that are available to them. Differences in structure and function mean different animals require different amounts of each nutrient.

In this topic, we are going to look at the functions of the different parts of each digestive system.

Monogastric digestive systems

Monogastric digestive systems are found in many mammals. Although there are some variations (ruminants and hindgut fermenters) the majority of monogastric animals have a very similar digestive system to humans.

In all animal species, digestion uses two processes:

- **Mechanical digestion** is the physical breakdown of food into smaller particles through mastication (chewing) and the muscular action of the digestive organs.
- **Chemical digestion** is the breakdown of foods using stomach acid (hydrochloric acid), enzymes and other digestive juices.

Carnivorous monogastric animals tend to have shorter intestines, whilst true herbivores tend to have the most adaptations (plant matter is much harder to digest due to the cellulose in plant cell walls).

Dentition

Teeth have evolved to enable animals to eat specific types of food:

- **Canines** are for holding prey and tearing flesh.
- **Incisors** are for cutting and tearing food.

- **Premolars** are for chewing and crushing food. In dogs, specially adapted premolars called **carnassial teeth** help to break down meat and bone.
- **Molars** are for crushing food.

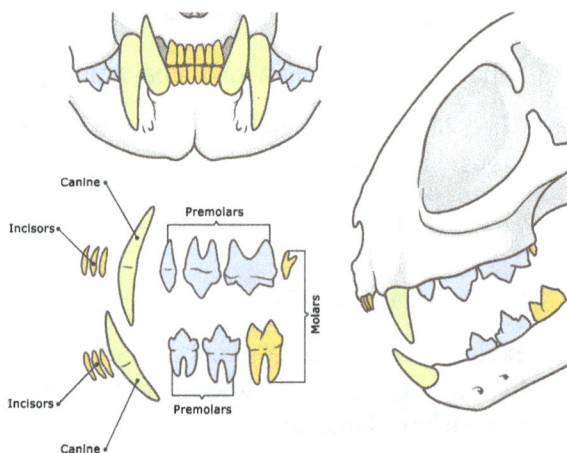

Carnivore teeth as seen in a domestic cat

Upper premolars
Upper molars
Lower premolars
Lower molars
Incisors

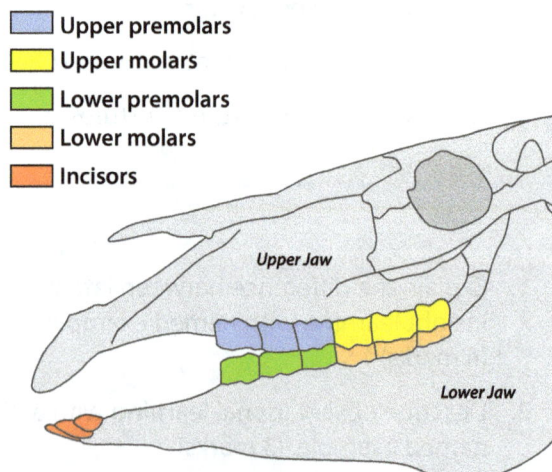

Herbivore teeth as seen in cattle

Organs of the monogastric digestive system

Structure	Function
Mouth	The mouth is the first part of the digestive system. Mastication takes place here as the first stage of mechanical digestion. Saliva adds the enzyme amylase, which begins the chemical digestion of starches.
Oesophagus	The muscular, tube-shaped organ which connects the mouth to the stomach.
Stomach	The stomach is a muscular organ which breaks down food using chemical and mechanical digestion. **Hydrochloric acid** (HCl) and enzymes called **proteases** and **lipases** break down the food whilst the muscular walls of the stomach churn the mixture into a substance called chyme.

Structure	Function
Small Intestine	The small intestine is divided into three parts: the **duodenum**, **jejunum** and **ileum**. In the duodenum, **bile** (made in the liver) is added from the gallbladder to neutralise the acidity of the chyme and emulsify fats. Additional digestive enzymes are added from the pancreas to continue the digestion of fats and proteins. In the **jejenum** and **ileum**, nutrients are absorbed into the bloodstream and sent to cells to be used or stored.
Large Intestine	The start of the large intestine is called the **caecum**. In carnivores, the caecum is very small and involved in the absorption of salts. It produces **mucus** to help with the passage of waste through the colon. In the colon, water and water-soluble vitamins are absorbed. **Microvilli** increase the surface area inside the large intestine, increasing the rate of absorption. Anything remaining is now waste material not required by the body. Waste material is stored in the rectum until it leaves the body via the anus.
Accessory organs	**Pancreas** – produces the enzymes **amylase** (starch), **protease** (proteins) and **lipase** (lipids) as well as the hormones **insulin** and **glucagon**. These hormones control blood sugar levels. **Gallbladder** – stores bile ready to be transported to the duodenum. **Salivary glands** – found in the mouth. Produce **amylase** to begin the breakdown of starches.

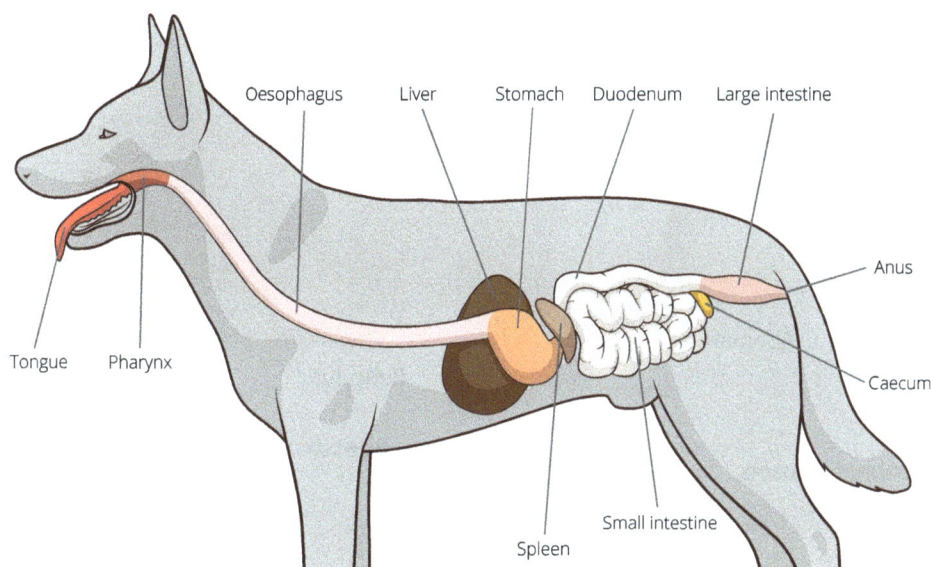

Monogastric digestive system found in dogs

Labels: Oesophagus, Liver, Stomach, Duodenum, Large intestine, Anus, Caecum, Small intestine, Spleen, Tongue, Pharynx

Important terms!

Monogastric: Digestive system with a single-chambered stomach. Example, found in cats and dogs.

Mastication: The act of chewing food.

Enzyme: Proteins which speed up chemical reactions.

Chyme: Partially digested food mixed with hydrochloric acid and digestive juices.

Hindgut fermenters

Plant matter is much harder to digest due to the structure of cellulose found in plant cell walls. To overcome this, true herbivores such as horses and rabbits have evolved an adaptation to their digestive systems called **hindgut fermentation**.

These species have an enlarged caecum at the start of their large intestines. Here, bacteria ferment the cellulose, producing a substance called **volatile fatty acids** which provide the animal with the majority of their energy requirements.

Rabbits are also known as **pseudo-ruminants**. In order to extract as much energy as possible from plant matter, rabbits produce nutrient-rich pellets called **caecotropes**. These contain nutrients which the rabbit has not yet extracted. So, rabbits consume caecotropes, which means they pass through the digestive system a second time.

Any matter that has been fully digested and is not required by the animal's body will leave the body as faeces.

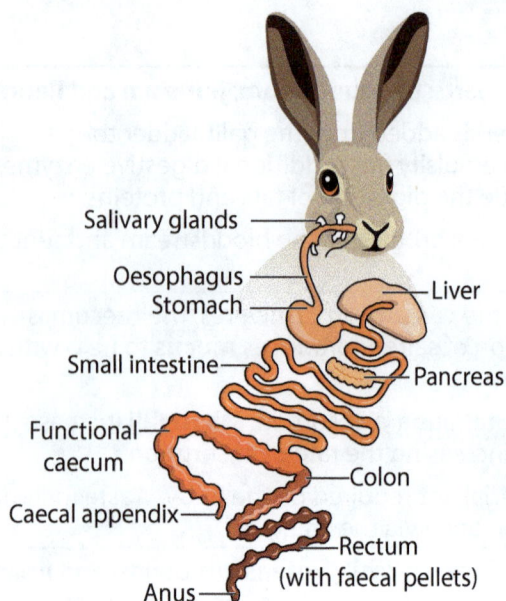

Salivary glands

Oesophagus
Stomach

Liver

Small intestine

Pancreas

Functional caecum

Colon

Caecal appendix

Rectum (with faecal pellets)

Anus

A rabbit's digestive system

Caecotropes

> **Important terms!**
>
> Hindgut Fermentation: A biological process in which bacteria ferment cellulose and other complex carbohydrates to extract nutrients from them.
>
> Caecum: A sack-like part of the digestive system where bacteria carry out fermentation in order to break down cellulose.
>
> Caecotrope: A nutrient-rich pellet which the rabbit must ingest in order to extract nutrients.

Ruminants

Another herbivore adaptation to consuming plant matter is **rumination**. This process is similar to hindgut fermentation in that it uses bacteria to ferment plant matter in order to extract nutrients. However, rather than this fermentation taking place in the hindgut, in ruminants it takes place in the foregut.

Ruminant animals such as cattle, goats and sheep have stomachs with four chambers: **rumen**, **reticulum**, **omasum** and **abomasum**.

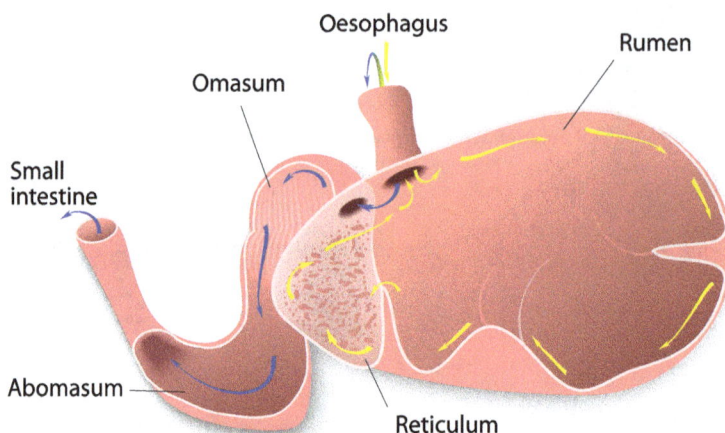

Oesophagus

Omasum

Rumen

Small intestine

Abomasum

Reticulum

Ruminant digestive system

Rumen	This is the largest of the four chambers in the ruminant stomach. Here, **bacteria ferment plant matter**, converting it into **volatile fatty acids**. Digested material leaves the rumen and passes into the reticulum.
Reticulum	In the reticulum, fully digested food passes through the **sieve-like structure** into the omasum.
	Anything which has not yet fully digested is formed into a **bolus** and **regurgitated so that the animal can chew it again**. Chewing (mechanical digestion) and amylase in saliva (chemical digestion) **break the bolus down,** ready to once again enter the rumen.
Omasum	The omasum has a layered structure, much like the pages of a book. This **increases the surface area** allowing water and salts to be absorbed.
Abomasum	The abomasum is often called the "true stomach" as it is most similar in structure to the monogastric stomach. Here, **hydrochloric acid** and **enzymes** are added to digest **proteins** and **fats** that the animal has consumed. From here, chyme passes to the duodenum.

Other digestive system adaptations

Birds

Birds have evolved a very different digestive system to mammals. The lack of teeth means they swallow food whole, using their tongue to push food to the back of the mouth.

- Food moves through the **oesophagus** to the **crop** (a temporary storage area).

- From here, food travels to the **proventriculus** where enzymes and hydrochloric acid are added to begin the digestive process.

- The food then moves into the **ventriculus** (also known as the gizzard). As birds do not have teeth to chew their food, they ingest small stones. The muscular walls of the ventriculus churn the food with the stones, grinding it down to make digestion more efficient.

- From the ventriculus, the food moves into the intestines. Birds have two **caeca** (the plural of caecum) where water absorption and bacterial fermentation help to further extract nutrients.

At the end of the digestive system, the **cloaca** is a muscular organ which connects to the reproductive and digestive systems. Eggs and waste matter leave the body from this point.

Birds have evolved different beak shapes depending on the food sources available to them. For example,… parrot beaks are very strong and designed to crack open the hard outer shell of nuts whilst nectar feeders, such as hummingbirds, have long thin beaks which can reach into flowers to extract the nectar.

Some species, such as pigeons, produce a crop milk to feed their offspring. The cells lining the crop secrete a nutrient rich substance which the birds regurgitate to feed their young.

Owls produce a pellet containing all non-digestible parts of their meal (such as bones) which they regurgitate.

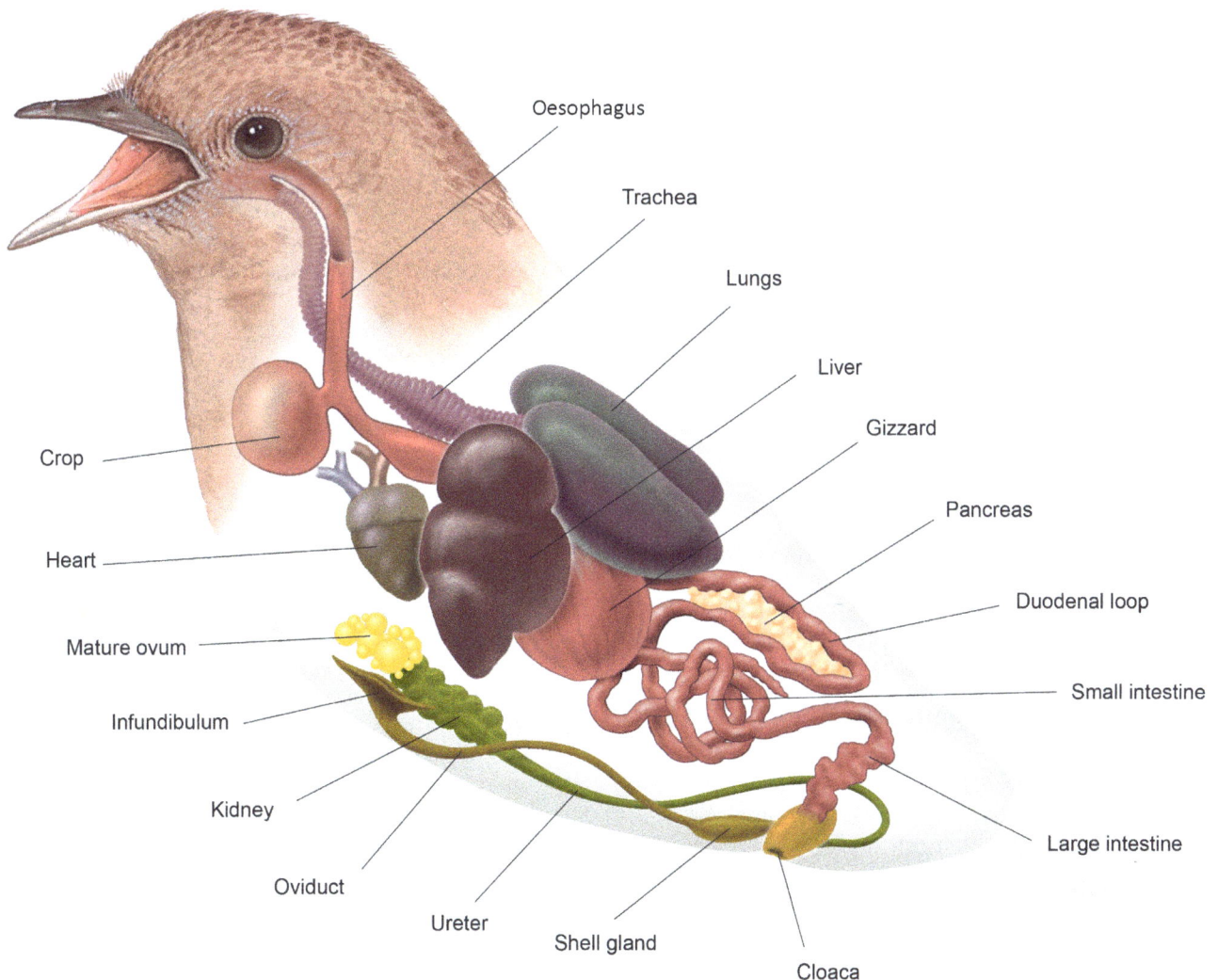

Bird digestive system

Antenna

Mandible

Eyes

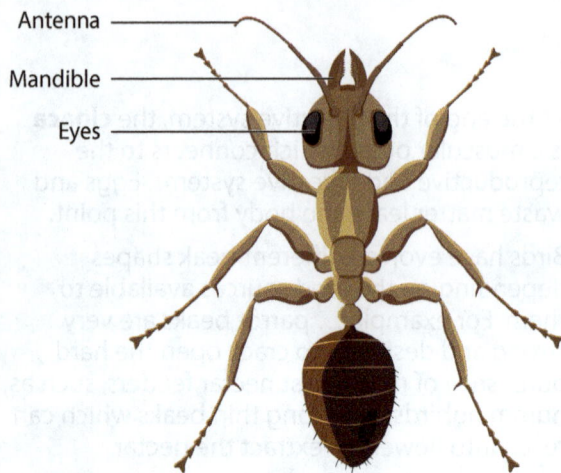

Mandibles on an ant

Insect Mouthpieces

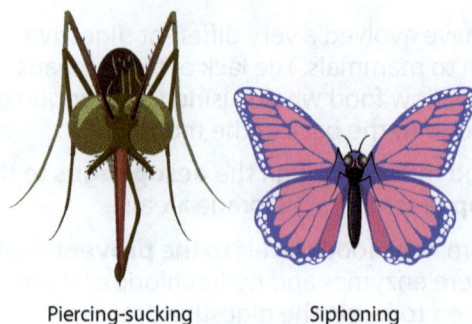

Piercing-sucking Siphoning

Variations of haustellate mouthparts

Insects

The invertebrate digestive system is very simple in comparison to vertebrate species. In insects, a single organ shaped like a coiled tube travels from mouth to anus. This is the **alimentary canal**. Digestion and absorption occur along this tube.

Insects have two key adaptations to their digestive systems:

- **Mandibles** are a pair of jaws for crushing or grinding food. They work with a side-to-side motion, not up and down as seen in mammals. Species that have mandibles are known as mandibulates. A good example of a species with mandibles is ants.

- Haustellate mouthparts are used for sucking liquid food into the mouth. Many species have evolved a **proboscis** which is used to pierce skin in order to drink blood (mosquitos) or to reach into flowers to extract nectar (butterflies).

Marine mammal dentition

Marine mammals such as whales and dolphins have evolved teeth that enable them to hunt successfully. They have huge variation in size, shape and design of teeth.

Toothed whales (**odontocetes**) have teeth comprised of three parts: the **crown**, the **root** and the **pulp**.

- The crown is the hard outer covering that protects the tooth.

- The root anchors the tooth into the jaw.

- The pulp contains the nerves and blood vessels.

Orcas have conical, interlocking teeth that can grip prey.

> **Important terms!**
>
> Gizzard: Another term for ventriculus.
>
> Mandibulates: Invertebrate species that ingest food using mandibles to chew or grind food.
>
> Haustellate: Invertebrate species that ingest food using a proboscis to suck liquid into their mouth.
>
> Keratin: A type of protein that forms nails, claws, scales and hair.

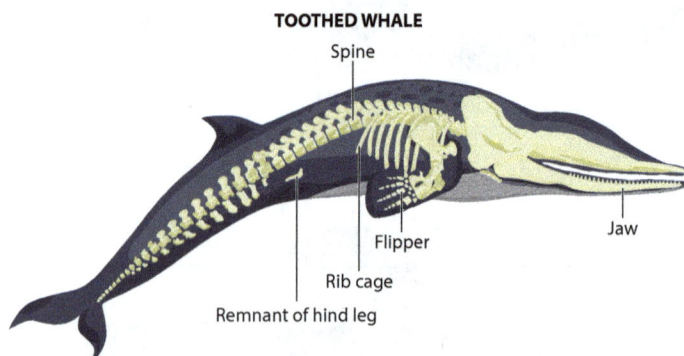

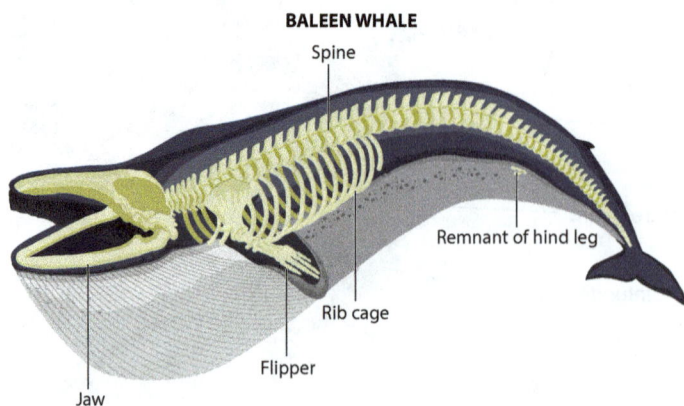

TOOTHED WHALE

Spine

Flipper

Jaw

Rib cage

Remnant of hind leg

BALEEN WHALE

Spine

Remnant of hind leg

Rib cage

Flipper

Jaw

The differences in mouth physiology between toothed whales and baleen whales

Baleen whales, such as the humpback, do not have teeth. Inside their mouth they have large **filter plates** that strain small prey such as **plankton** from the water. These plates are made of keratin and have a hair-like structure to them.

Reptiles and amphibians

Snakes

Features of a snake's digestive system include:

- Short digestive tract adapted for digesting meat.

- Powerful stomach acid to break down an entire carcass, including bones.

- Backwards-facing teeth to prevent prey escaping. They do not use their teeth for chewing but rather to swallow prey whole. Snakes have highly flexible jaws that unhinge, allowing them to swallow very large prey.

Tortoises

Tortoises have long intestines with a large caecum to digest fibrous plant material. These species use hindgut fermentation, similar to rabbits and horses.

Tortoises have evolved a beak-like structure

Bearded dragons

These have a medium-length digestive tract, adapted to digest both plants and meat. This adaptation is found in many omnivorous lizards.

Lizards have solid jaws evolved for chewing plant matter, meat and invertebrates.

Recap Questions

1. Name three digestive enzymes and the type of food they break down.

2. Where is water absorbed in the digestive system?

3. Why do carnivores have shorter intestines than herbivores?

4. Why do some herbivores use hindgut fermentation as part of their digestive processes?

5. How are caecotropes different from faeces?

6. Name three ruminant animals.

7. Explain the function of each chamber of the ruminant stomach.

8. Why do ruminant animals regurgitate food?

9. Why do birds ingest stones?

10. What is the function of the cloaca?

11. Explain the difference between mandibulates and haustellates. Give an example of each.

12. Why do owls produce pellets?

6.2 The structure and function of the respiratory system in relation to animal physiology

The **respiratory system** allows air to enter and exit the body. Gas exchange is essential to ensure cells receive oxygen and remove carbon dioxide. In many species, the respiratory system also plays a vital role in thermoregulation.

The respiratory system links very closely with the circulatory system (see section 6.3).

Mammalian respiratory system

During ventilation, the **diaphragm** contracts and the intercostal muscles **pull the ribcage outwards**. This reduces the pressure inside the thorax (chest) causing air to flow into the lungs, inflating them.

When the diaphragm relaxes, the intercostal muscles pull the ribcage inwards **increasing the pressure in the thorax**, forcing the air out of the lungs.

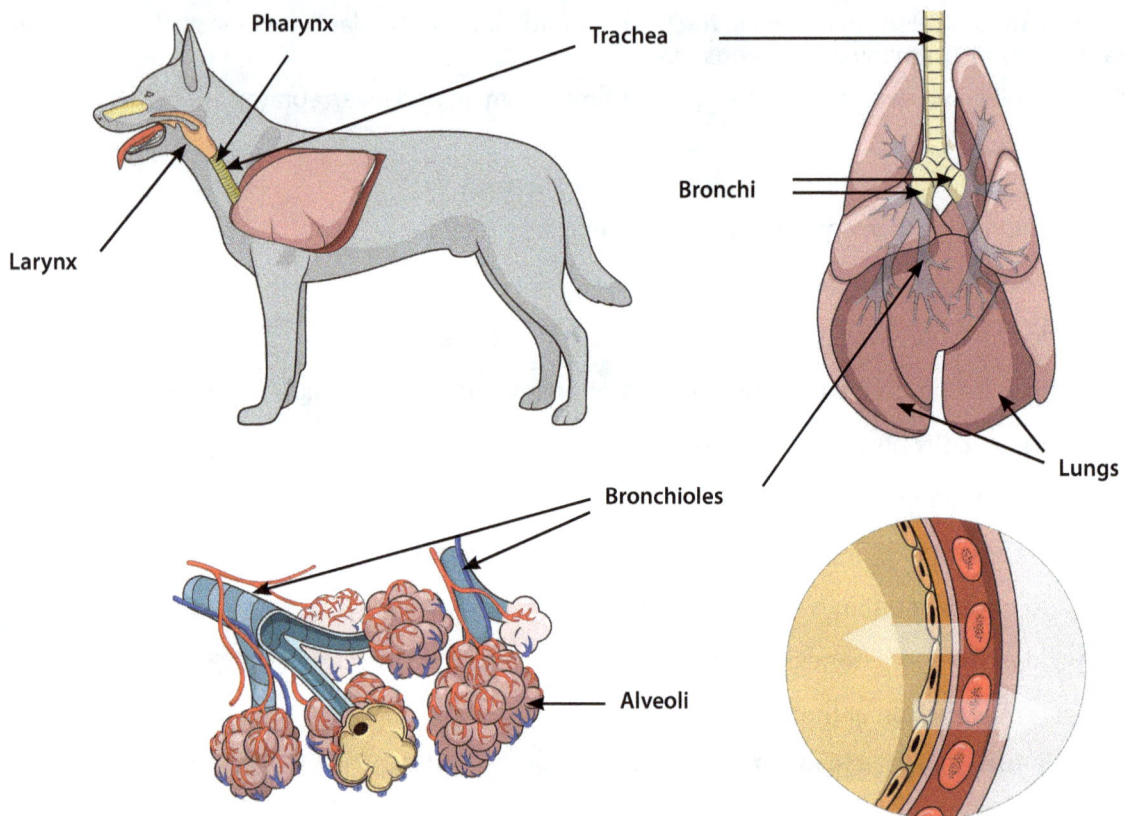

A dog's respiratory system

Larynx – also known as the voice box. The larynx is responsible for vocalisation. It also houses the **epiglottis**, a flap of skin which blocks the entrance to the trachea when the animal swallows to prevent food or water entering the lungs.

Pharynx – The area of the throat which connects the nose and mouth to the trachea.

Trachea – a tube which connects the larynx to the lungs. Rings of cartilage prevent the trachea from collapsing.

Bronchi – the trachea splits into two airways called bronchi, one for each lung.

Bronchioles – The bronchi divide into smaller airways called bronchioles. These deliver air to the alveoli.

Alveoli – Tiny balloon-like structures in the lungs which inflate and deflate as air enters and leaves them. These are the site of gas exchange.

Lungs – the main organ of the respiratory system. Filled with a branch-like system of bronchioles and alveoli.

Gas exchange

In mammals, the alveoli share a membrane with tiny capillaries. Gases are able to diffuse through the shared membrane.

1. Oxygen diffuses into the blood to be transported around the body.

2. Carbon dioxide diffuses out of the blood into the alveoli to be breathed out of the body.

In the blood, oxygen is transported by a protein in red blood cells called haemoglobin.

- Oxygen binds with the haemoglobin to form oxyhaemoglobin.

- When oxygen is delivered to muscle cells, it binds to a protein called myoglobin. It is stored here until the muscle needs it during cellular respiration.

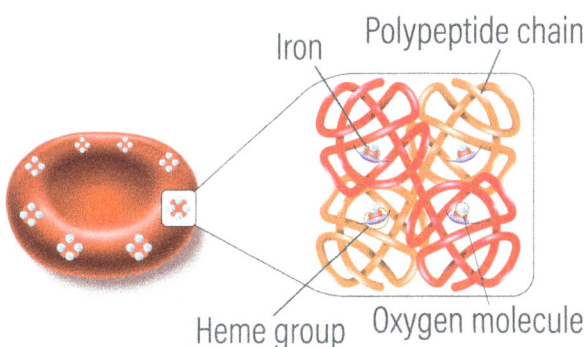

Gas exchange

Capillary · Wall of alveoli · Oxygen-poor blood · Oxygen molecule · Red blood cell · Carbon dioxide molecule · Alveoli · Oxygen-rich blood · CO2 · O2

Iron · Polypeptide chain · Heme group · Oxygen molecule

Haemoglobin

Important terms!

Gas exchange: The diffusion of gases between the alveoli and blood.

Ventilation: Another term for breathing.

Thermoregulation: Methods that the body uses to control internal body temperature.

Thorax: The area of the body inside the ribcage.

Haemoglobin: The protein in red blood cells which transports oxygen.

Oxyhaemoglobin: The oxygenated form of haemoglobin.

Myoglobin: A protein found in muscle cells which stores oxygen for use in cellular respiration.

Respiration: The process in which glucose is converted into energy.

Bird respiratory system

The bird respiratory system is very different from the mammal system. Birds do not have lungs that inflate and deflate as they breathe. Instead, their body has nine air sacs that inflate then push the air through air capillaries where gas exchange takes place.

Birds also have specialised bones called **pneumatic bones**. They have a hollow, honeycomb-like structure which acts as an extension to the air sacs. This improves their ability to breathe at high altitudes where there is less oxygen.

Birds do not have a diaphragm. Instead, the intercostal and abdominal muscles expand and contract the chest, inflating and deflating the air sacs.

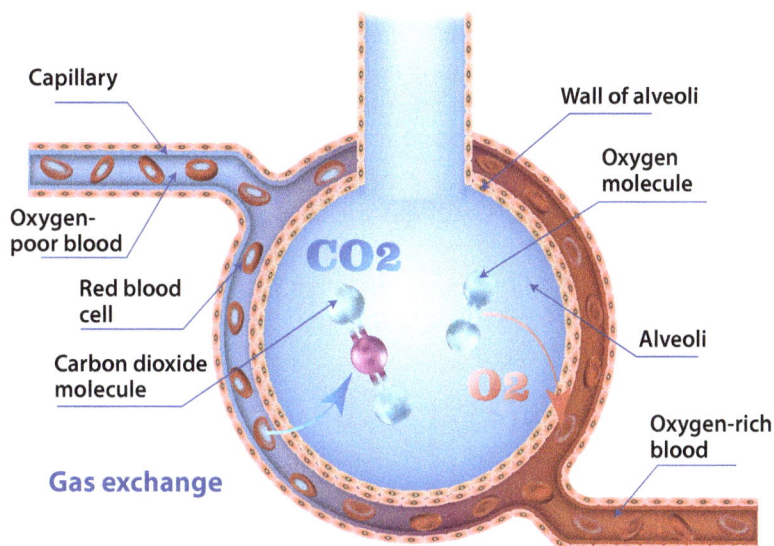

Pneumatic bone

Nares – the bird equivalent of nostrils.

Air sacs – 7 or 9 air sacs which move air through the respiratory system in one direction. These extend into specialised bones called pneumatic bones.

Air capillaries – the site of gas exchange in birds.

Larynx – not used for vocalisation in birds. The larynx prevents food or water from entering the respiratory system.

Syrinx – found where the trachea splits into two bronchi and is used for vocalisation.

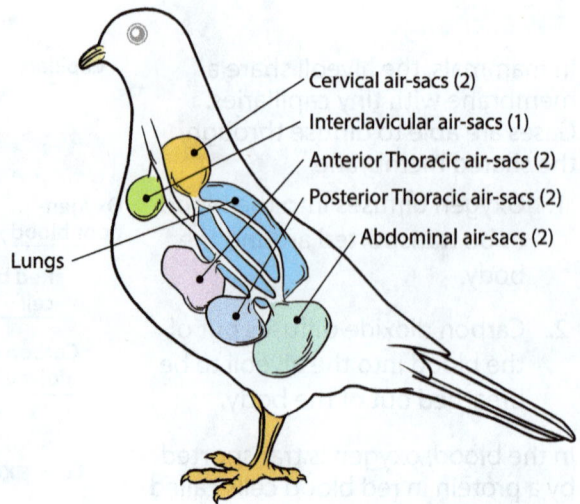

- Cervical air-sacs (2)
- Interclavicular air-sacs (1)
- Anterior Thoracic air-sacs (2)
- Posterior Thoracic air-sacs (2)
- Abdominal air-sacs (2)

Lungs

A bird's respiratory system

Birds need two respiratory cycles for air to travel fully through the respiratory system.

1. Inhalation: air passes through the trachea into the posterior air sacs.

2. Exhalation: air moves from the posterior air sacs into the air capillaries of the lungs.

3. Inhalation: air moves from the lungs into the cranial air sacs.

4. Exhalation: air moves from the cranial air sacs, through the trachea and out of the body.

It is important to remember that Inhale 1 and Inhale 3 are happening at the same time and Exhale 2 and Exhale 4 are happening **at the same time**. Birds never exhale all air from their body in the same way mammals do.

Important terms!

Pneumatic Bones: Bones with a hollow honey-comb structure.

Posterior: Towards the back of the body.

Cranial: Towards the skull.

Unidirectional: Flowing in one direction.

Unlike mammals, birds are able to extract oxygen from the air during both inhalation and exhalation. This is due to the **unidirectional flow of air** through the air capillaries. This makes a bird's respiratory system **very efficient** at gas exchange.

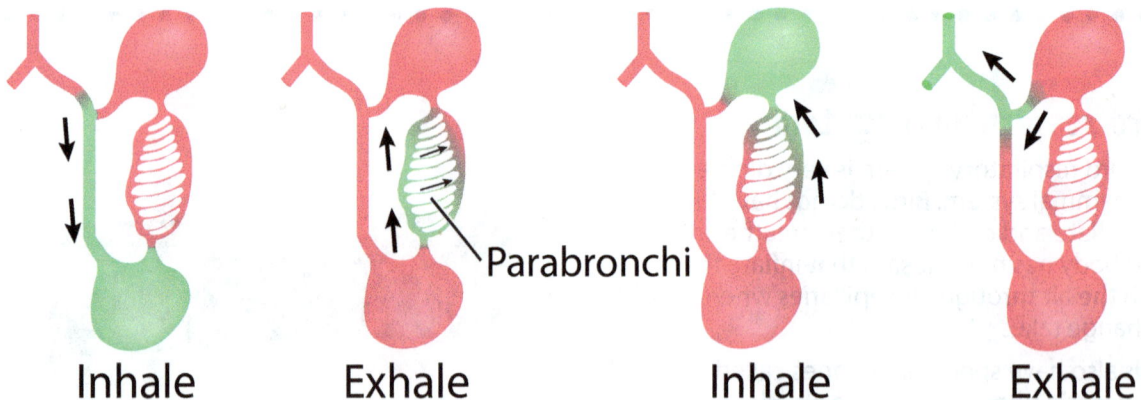

Inhale   Exhale   Parabronchi   Inhale   Exhale

Bird respiration

Fish respiratory system

Unlike mammals and birds, fish use gills to extract oxygen from water.

- Water passes through the fish's mouth and over the gills.
- **Dissolved oxygen in the water** diffuses into the capillaries whilst carbon dioxide diffuses out.
- The water then leaves the body through the **operculum** or gill cover.

The gills are large in comparison to body size with many small filaments, which gives a huge surface area for gas exchange to take place.

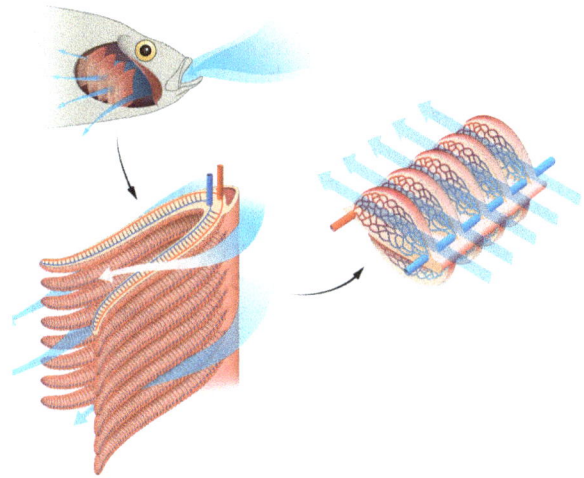

Respiration in fish

• •

Reptile and amphibian respiratory system

Reptiles breathe using lungs in a similar way to mammals. Unlike mammals, most reptiles do not have a diaphragm. Instead, they change the volume of their chest cavity using the intercostal muscles of the ribs and the muscles of the abdomens. Many reptiles cannot breathe if they lay on their back.

Like reptiles, amphibians do not have a diaphragm. Instead, they use a pouch of skin in their throat called a **buccal cavity** to push air into the lungs for gas exchange.

Unlike reptiles, amphibians have very thin skin which is able to diffuse oxygen from water directly into the blood supply. This is only possible if their skin is wet, which is why reptiles must remain moist at all times.

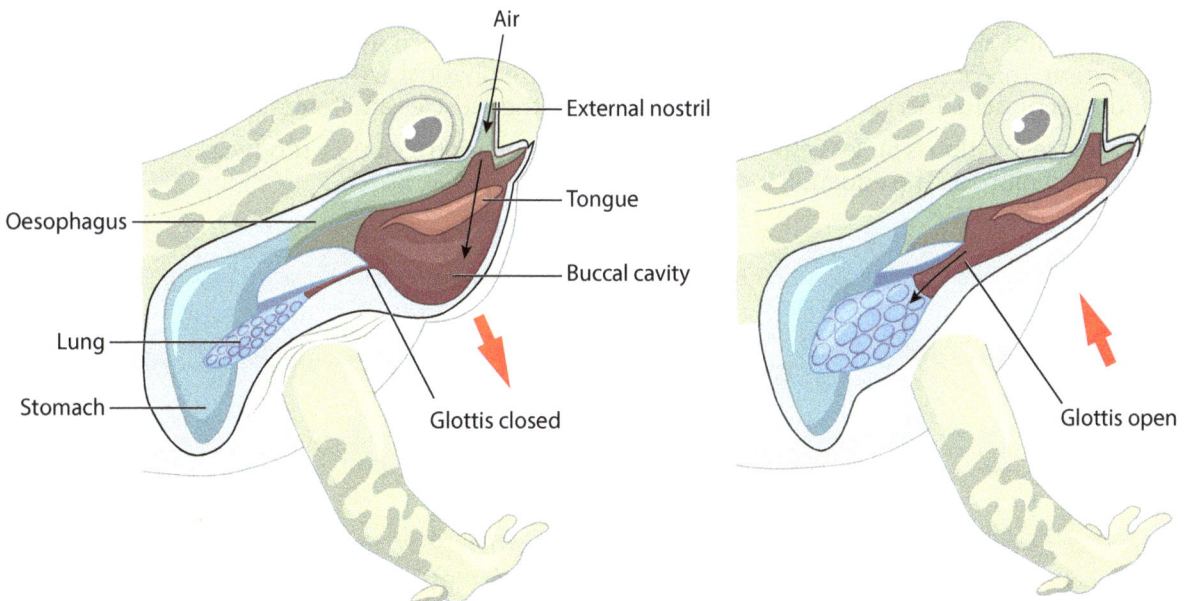

Respiration in frogs

Invertebrate respiratory system

Many invertebrates, such as insects and arachnids, have evolved a highly efficient respiratory system.

Air enters the body through small openings on the sides of the body called **spiracles**. The air then travels through a network of tubes called **tracheae,** which delivers the air to body tissues. Gas exchange occurs directly between the tracheae and body tissues, bypassing blood entirely.

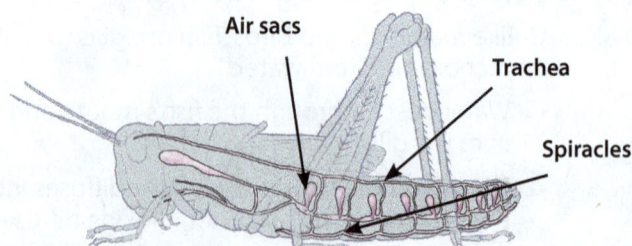

Respiratory system of an insect

Respiration

During cellular respiration, cells convert glucose into a molecule called **adenosine triphosphate** or **ATP**. This molecule is used to power the processes that take place throughout the body.

> **Remember!**
> Note: respiration **is not** the same as breathing.

Aerobic respiration

Aerobic respiration uses oxygen to break down glucose molecules to produce ATP. This uses a series of complex processes which take place in the **cytoplasm** and **mitochondria**.

Aerobic respiration can produce up to 38 ATP per molecule of glucose. The equation for this is:

$$C_6H_{12}O_6 + 6O_2 \longrightarrow 6CO_2 + 6H_2O + \text{(Energy)}$$

$$\text{Glucose} + \text{Oxygen} \longrightarrow \text{Carbon dioxide} + \text{water} + \text{(Energy)}$$

Anaerobic respiration

If oxygen is not available, the animal body will continue to produce ATP using **anaerobic respiration**. This occurs in the muscle cells during intense physical activity and produces the waste product **lactic acid**. This causes a burning sensation in the muscles and forces the animal to stop.

Anaerobic respiration is far less efficient than aerobic respiration as it only produces 2 ATP per molecule of glucose. The equation for this is:

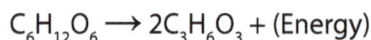

$$C_6H_{12}O_6 \longrightarrow 2C_3H_6O_3 + \text{(Energy)}$$

$$\text{Glucose} \longrightarrow \text{Lactic Acid} + \text{(Energy)}$$

Recap Questions

1. Give three differences between aerobic and anaerobic respiration.

2. What is the function of the buccal cavity in frogs?

3. How does frog skin contribute to gas exchange?

4. Why can animals not rely on anaerobic respiration alone?

5. Explain how the insect respiratory system is different from mammals.

6. Why do birds have pneumatic bones?

7. Describe the process of ventilation in birds.

8. What adaptations does the bird respiratory system have that make it more efficient than mammals?

6 Anatomy and physiology

6.3 The structure and function of the circulatory system in relation to animal physiology

The circulatory system transports blood, oxygen and nutrients to all cells in the body, and transports waste products away from them.

Structure and function of the mammalian circulatory system

The mammalian circulatory system is a **closed double circulatory system**, meaning blood passes through the heart twice in one complete circuit. This ensures efficient oxygen delivery and waste removal.

Blood

Blood has four main components:

- Plasma
- Erythrocytes (red blood cells)
- Leukocytes (white blood cells)
- Platelets.

Plasma

Plasma is a pale yellow liquid that makes up approximately 55% of blood. It is composed of water, proteins (such as albumin, globulin and fibrinogen), electrolytes, nutrients, hormones and waste products. Plasma has several vital functions, including transporting nutrients and hormones, removing waste products, maintaining osmotic pressure and pH balance.

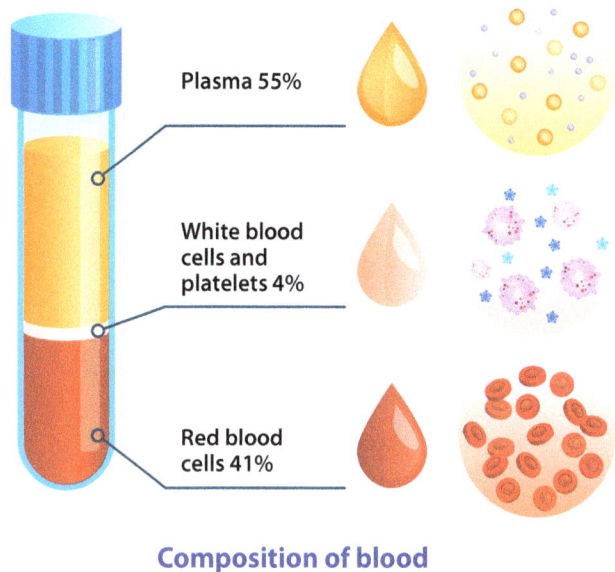

Plasma 55%

White blood cells and platelets 4%

Red blood cells 41%

Composition of blood

Erythrocytes

Erythrocytes are specialised blood cells which contain **haemoglobin**. This is a specialist protein which binds to oxygen in the lungs and releases it into body tissues. Red blood cells are produced in the bone marrow and have a lifespan of approximately 120 days. At the end of their lifespan, they are broken down in the spleen and recycled in the liver.

Leukocytes

Leukocytes are key components of the immune system and defend the body against infection from pathogens. Each type of cell has its own role:

- Neutrophils are the first responders to bacterial and fungal infections. They perform phagocytosis, engulfing and destroying pathogens.

> **Important terms!**
> Erythrocytes: Red blood cells
> Leukocytes: White blood cells
> Plasma: The liquid portion of blood.
> Thrombocytes: Also called platelets, these are cells that form a clot when a blood vessel is damaged.

- Eosinophils and basophils combat parasitic infections, particularly helminths (worms). They also play a role in allergic reactions by releasing histamine to bring more blood into the affected area.

- Monocytes are the largest of the white blood cells. Their role in immunity is phagocytosis. This means they engulf larger pathogens or debris and digest it in order to destroy it.

- There are also T-lymphocytes and B-lymphocytes. See the section on the lymphatic system for more about them.

Platelets

Platelets (thrombocytes) are fragments of cells that form clots when a blood vessel is damaged. Clots prevent further blood loss and reduce the risk of infection.

The heart

The heart is a muscular organ formed from cardiac muscle that pumps blood through the circulatory system. In mammals, the heart has four chambers:

- **Right Atrium**: Receives deoxygenated blood from the body via the **vena cava**.
- **Right Ventricle**: Pumps deoxygenated blood to the lungs via the **pulmonary artery**.
- **Left Atrium**: Receives oxygenated blood from the lungs via the **pulmonary vein**.
- **Left Ventricle**: Pumps oxygenated blood to the body via the **aorta**.

Circulatory system

Mammals have a **double circulatory system**. This means that each side of the heart is pumping blood in separate directions (to the lungs and around the body) at the same time.

In order to facilitate this, the muscular walls of the left ventricle are much thicker than those of the right.

- This allows the right ventricle to pump blood at a lower pressure to the lungs, protecting the delicate capillaries that receive oxygen from the alveoli.
- At the same time, the thicker walls of the left ventricle allow oxygenated blood to be pumped with a high enough pressure around the rest of the body. This is a very efficient system and allows for sufficient oxygen to reach all body tissues.

Valves in the heart prevent the backflow of blood from a ventricle to an atrium. If backflow were to happen, oxygen delivery would be severely affected, making the animal feel very tired and unable to perform all necessary actions to survive.

- The **bicuspid valve** is between the left atrium and the left ventricle.
- The **tricuspid valve** is between the right atrium and the right ventricle.

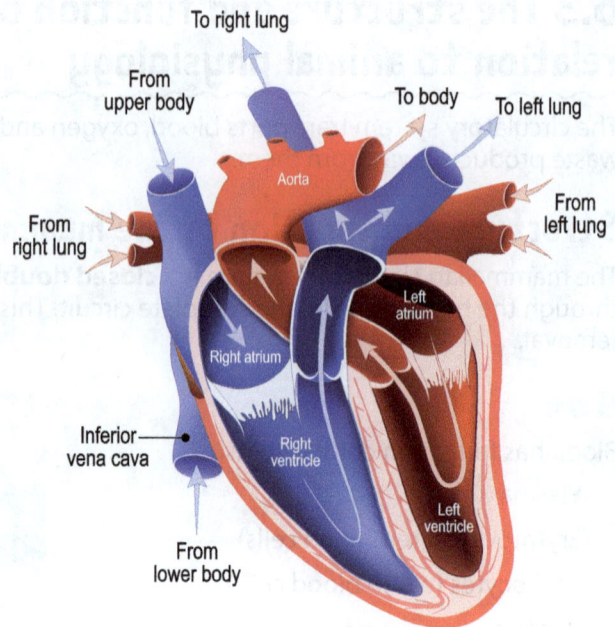

Structure of the heart

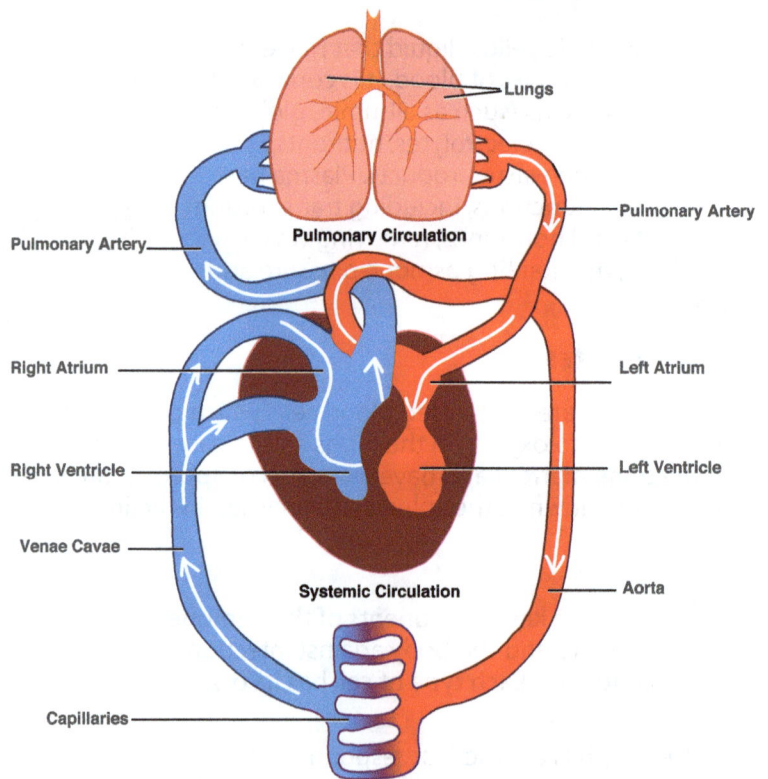

Veins and arteries connected to the heart

Valves are supported by **chordae tendineae**, string-like structures which prevent inversion during contraction.

Electrical conduction system

The heart is a unique muscle because it has its own electrical conduction system, which controls the rate at which the heart contracts. In order to pump blood effectively, the chambers of the heart need to contract in a specific order.

- The **sinoatrial node**, also known as the pacemaker, generates electrical impulses at regular intervals. These impulses cause the atria (upper chambers) to contract, forcing blood into the ventricles.

- The **atrioventricular node** (**AV node**) is located at the base of the right atrium. It slows the impulse to ensure the ventricles do not contract until the atria have fully contracted.

- **Purkinje fibres** distribute the electrical signal throughout the muscular walls of the ventricles, ensuring the muscle contracts from the apex (bottom) upwards, forcing the blood through the pulmonary artery and the aorta.

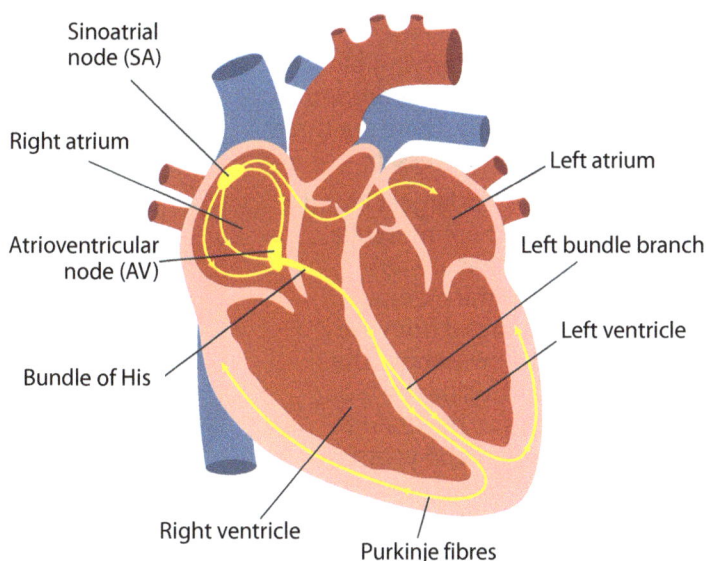

Labels: Sinoatrial node (SA), Right atrium, Atrioventricular node (AV), Bundle of His, Right ventricle, Left atrium, Left bundle branch, Left ventricle, Purkinje fibres

Electrical conduction system of the heart

- The **Bundle of His** conducts the electrical signals from the AV node to the ventricles. It splits into two branches to ensure that both ventricles receive the signal simultaneously.

Heart rate regulation

The rate at which the sinoatrial node produces the electrical signal is determined by several factors within the body. As well as the following factors, the heart rate impulse depends on the species (smaller animals tend to have a faster heart rate)

Environmental temperature

When an animal is too hot, the heart rate will increase along with vasodilation of the blood vessels in the skin. This directs more blood to the surface of the body, away from the core, which reduces body temperature. (There are also other mechanisms to reduce body temperature, such as sweating).

Exercise

During physical exertion, an increase in carbon dioxide levels in the blood triggers an increase in heart and breathing rate, to bring more oxygen into the body.

Fight or flight response

When the animal is threatened, the sympathetic nervous system triggers the **fight-or-flight response**. In the fight-or-flight response, increased levels of adrenaline causes a rapid increase in heart rate, to increase blood flow to the skeletal muscles. This makes the animal faster and stronger so it can escape predators (flight) or fight back.

> **Important terms!**
>
> Sinoatrial node: The pacemaker that creates the electrical impulse, causing the heart to beat.
>
> Atrioventricular node: Slows the electrical impulse created by the sinoatrial node, ensuring the ventricles contract after the atria.
>
> Vasodilation: Widening of blood vessels
>
> Sympathetic nervous system: A system that controls the 'flight or fight' response. It is not under conscious control.
>
> Adrenaline: A hormone that targets organs in the body, to get it ready for immediate action. It controls various actions to ensure as much energy as possible is available for the muscles.

Blood vessels

Blood vessels form a closed circulatory system carrying blood from the heart and lungs around the rest of the body. There are three main types of blood vessel:

- **Arteries** are the largest blood vessels with thick, muscular, elastic walls that are able to withstand the pressure of blood pumped from the heart. A narrow **lumen** (the tube formed inside the blood vessel) helps to maintain a high enough blood pressure to transport blood around the entire body. All arteries carry **oxygenated blood** except for **the pulmonary artery**.

- **Veins** have much thinner walls than arteries, as the pressure within them is lower. Valves along their length prevent the backflow of blood. **All veins carry deoxygenated blood** except for **the pulmonary vein**.

- **Capillaries** are microscopic vessels with walls that are only one cell thick. This allows for efficient diffusion of gases, nutrients and waste between blood and body tissues.

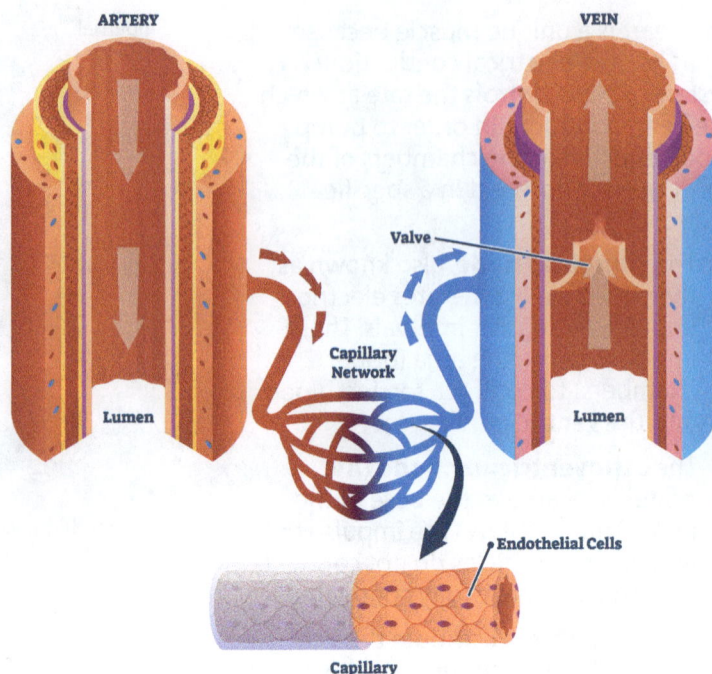

ARTERY • VEIN • Valve • Capillary Network • Lumen • Lumen • Endothelial Cells • Capillary

Blood vessels

The structure and function of the lymphatic system

The **lymphatic system** is an open circulatory system consisting of vessels, lymph nodes and organs that work alongside the circulatory system of the heart. Its primary functions are fluid balance, immune defence, aiding in fat digestion and transport.

Fluid balance

Excess interstitial fluid is collected from tissues and returned to the bloodstream, preventing oedema (swelling).

Fat digestion

Lipids are transported from the small intestines via **lacteals** (lymphatic capillaries) and delivered to the bloodstream for delivery around the body.

Transport

Lymph (the fluid within the lymphatic system) consists of water and electrolytes, immune system cells, proteins such as antibodies, fats and waste materials. The lymphatic system plays a vital role in transporting these around the body.

Lymphatic system structure

Lymphatic vessels are similar in structure to blood vessels but with thinner walls and more valves. Unlike the circulatory system, the lymphatic system does not have a pump pushing fluid around. Instead, it relies on:

- skeletal muscle contractions
- valves to prevent the backflow of blood
- changes in thoracic pressure during breathing.

Lymph nodes are small structures located around the animal body along the lymphatic vessels. Their role is to filter lymph, trapping pathogens, cell debris and cancer cells. They provide a site for immune cells to recognise and respond to threats.

The **thymus gland** is a small gland located near to the heart. It produces and matures **T-cells** which are essential for adaptive immunity. The thymus gland is mostly active in juvenile animals and shrinks when the animal reaches adulthood.

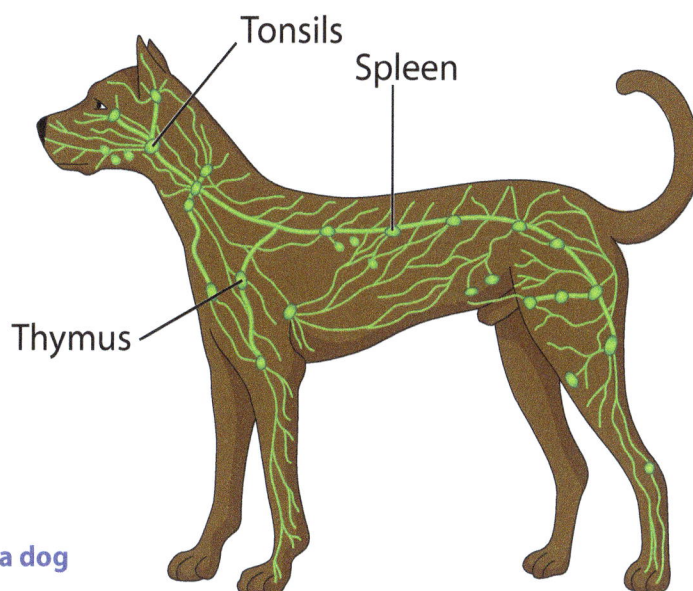

Lymphatic system of a dog

The **spleen** filters blood, removing and breaking down old and damaged red blood cells and pathogens. It stores white blood cells and platelets for when they are needed.

Tonsils are located in the throat and act as the first line of defence against inhaled or ingested pathogens.

Bone marrow is found in the interior of bones. It produces all blood cells, including lymphocytes (B-cells).

Peyer's patches are clusters of lymphoid tissue found in the intestines. They monitor and respond to pathogens in the digestive tract.

T-Lymphocytes (**T-cells**) are involved in **cell-mediated immunity** – which means they destroy infected or abnormal cells. **T**-cells mature in the **t**hymus gland.

B-Lymphocytes (**B-cells**) play a key role in **humoral immunity** – which means they produce antibodies to neutralise pathogens.

There are two types of B-cells:

- plasma cells produce antibodies
- memory cells hold details of pathogens for long-term immunity.

B-cells mature in **b**one marrow.

> **Important terms!**
>
> Open circulatory system: A circulatory system where fluid can enter or leave the surrounding tissues.
>
> Interstitial fluid: Fluid that leaks from blood capillaries into tissues
>
> Thoracic pressure: The pressure within the chest cavity.

• •

Comparison of different taxa

Aves

Like mammals, aves (birds) have a **four-chambered heart** and **double circulatory system**. This ensures their high demand for oxygen during flight can be met.

Bird erythrocytes are elliptical, giving a greater surface area and unlike mammals, bird red blood cells have a nucleus. This means that these cells can perform protein synthesis, repairing cellular damage and therefore lasting longer than in mammals.

Pisces

Unlike mammals and birds, pisces (fish) have a **two-chambered heart** consisting of one atrium and one ventricle. This creates a single circulatory system with the blood flowing in one continuous loop through the heart, gills and body. This limits the pressure at which the blood can be pumped as the gill capillaries are very delicate.

Reptilia and amphibia

Reptiles and amphibians have an **adapted double circulatory system**. Amphibians and most reptiles have a **three-chambered heart** (two atria, one ventricle) which allows partial mixing of oxygenated and deoxygenated blood. This is less efficient for oxygen delivery but works well in ectothermic species, where the demand for oxygen is lower.

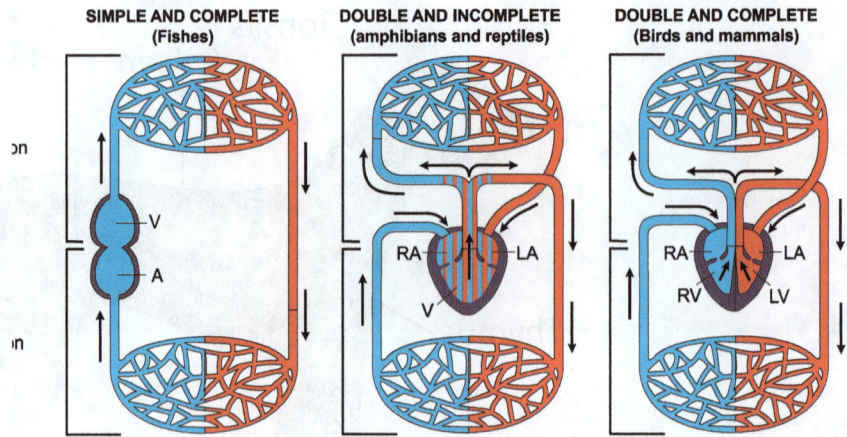

SIMPLE AND COMPLETE (Fishes)

DOUBLE AND INCOMPLETE (amphibians and reptiles)

DOUBLE AND COMPLETE (Birds and mammals)

Blood circulation systems in different taxa

Invertebrates

Invertebrates have an **open circulatory system** in which blood (haemolymph) surrounds organs and is not confined to vessels. This is a low-pressure system suited for smaller organisms with lower metabolic needs.

> **Important terms!**
>
> Aves: The vertebrate class which comprises birds.
>
> Pisces: The vertebrate class which comprises fish.
>
> Haemolymph: The fluid found within the body of invertebrates that has properties of both blood and lymphatic fluid.

Recap Questions

1. What is the role of erythrocytes?

2. What is the name of specialist protein within an erythrocyte? What is its function?

3. Name three leukocytes and give their functions.

4. What is the role of the atria within the heart?

5. Where does the pulmonary vein transport blood to and from?

6. Why does the left ventricle have thicker walls than the right ventricle?

7. Explain the steps that produce a heart beat.

8. Give two factors that affect heart rate and explain how they affect it.

9. Which blood vessels have the thickest walls? Why?

10. Which blood vessels have valves? Why?

11. Give three functions of the lymphatic system.

12. The lymphatic system does not have a pump. How does lymphatic fluid move through the system?

13. Name three organs or glands that form the lymphatic system. What are their functions?

14. Fish have a two chambered heart. Why would this not work in other species of vertebrates?

15. Why is a double circulatory system more efficient than a single circulatory system?

6.4 The structure and function of the endocrine system in relation to animal physiology

The animal body is a series of systems that must function together in order to ensure the animal is healthy. The endocrine system is one of the main communication networks in the body, working alongside the nervous system to ensure that all systems are coordinated. The **endocrine system** produces **hormones**, chemical messengers that allow the body to maintain homeostasis or balance.

Hormones are chemcials that each have a specific function in the body. They are released in response to specific triggers and cause the body to carry out actions to ensure homeostasis is maintained.

Major endocrine glands and their functions

Hypothalamus

The **hypothalamus** is located at the base of the brain above the pituitary gland. It receives signals from the nervous system which it cascades to the pituitary gland.

The hypothalamus is responsible for regulating many key processes in the body such as thermoregulation, thirst, hunger and sleep. It produces hormones that control the pituitary gland.

Pituitary gland

The **pituitary gland** is found beneath the hypothalamus in the brain. It releases hormones that regulate other endocrine glands.

The **anterior pituitary gland** produces hormones such as **growth hormone** (GH), **prolactin** and **adrenocorticotropic hormone** (ACTH).

- **Prolactin** plays a major role in reproduction, lactation, parental behaviour, immune system activity and in some species water and salt balance.

- **Adrenocorticotropic hormone** is responsible for regulating cortisol production which influences metabolism, immunity and stress responses

The **posterior pituitary gland** stores and releases hormones such as **oxytocin** and **antidiuretic hormone** (ADH).

- **Oxytocin** stimulates uterine contractions during parturition (giving birth), allows the mother to bond with her offspring and plays a role in herd or pack behaviours in social mammals such as wolves, primates and elephants.

- **Antidiuretic hormone** regulates osmosis (water balance in the body) by controlling the amount of water reabsorbed by the kidneys.

Pancreas (endocrine Function)

The **pancreas** is located near to the liver and spleen. It has both endocrine (hormone) and exocrine (digestive) tissue. The endocrine portion of the pancreas produces the hormones insulin and glucagon which regulate blood sugar levels.

- **Insulin** lowers blood sugar by instructing body cells to absorb sugar and instructing the liver to convert sugar into **glycogen** for storage.

- **Glucagon** increases blood sugar by triggering the breakdown of glycogen to release the stored glucose back into the blood.

Adrenal glands

The **adrenal glands** are located on top of the kidneys. Their function is to produce hormones involved in stress response and metabolism. The adrenal cortex produces cortisol and aldosterone, whilst the adrenal medulla produces adrenalin and noradrenalin.

- **Cortisol** (also known as the stress hormone) helps the body to cope with periods of stress by increasing blood glucose, blood pressure and alertness. It also plays a role in metabolism and immune system regulation.

- **Aldosterone** is a steroid hormone that plays a role in maintaining salt and water balance in the animal body. This enables the body to maintain blood pressure and blood pH.

- **Adrenaline** plays a key role in the "fight or flight" response. Adrenaline increases heart rate and blood flow, dilates the airways allowing more oxygen to be absorbed, increases blood glucose levels and redirects blood away from non-essential areas such as the digestive system to the skeletal muscles.

- **Noradrenaline** is a hormone and neurotransmitter. It also plays a key role in the "fight or flight" response. Noradrenaline causes **vasoconstriction**, a narrowing of the blood vessels which increases blood pressure. It increases the heart rate and prepares the muscles for action.

Thyroid and parathyroid glands

The thyroid and **parathyroid glands** are located in the neck. Their function is to produce hormones that regulate metabolism, growth and development.

- Thyroxine (**T4**) plays several major roles in metabolism and growth. It controls the rate at which cells use oxygen and energy, helps regulate metabolic rate, stimulates heat production in the cells of endothermic species and enhances protein synthesis. Thyroxine is converted into its active form triiodothyronine in the body tissues in order to carry out these functions.
- Triiodothyronine (**T3**) is the active form of thyroxine.
- **Parathyroid hormone** regulates calcium levels in the blood and bones.

Ovaries

Ovaries are found in the pelvic region in female animals. They produce the hormones oestrogen and progesterone which play a major role in reproduction and behaviour.

- **Oestrogen** is a steroid hormone responsible for ovulation (development and release of ova), promotes female mating behaviours, maintains bone density and cardiovascular health.
- **Progesterone** prepares the uterus for implantation of a fertilised egg then maintains the conditions to support the foetus throughout the pregnancy.

Testes

Testes are located in the scrotum in male animals. They produce **testosterone** which regulates sperm production and influence behaviour.

Testosterone regulates the production of sperm in the testes. It is also responsible for male mating behaviours.

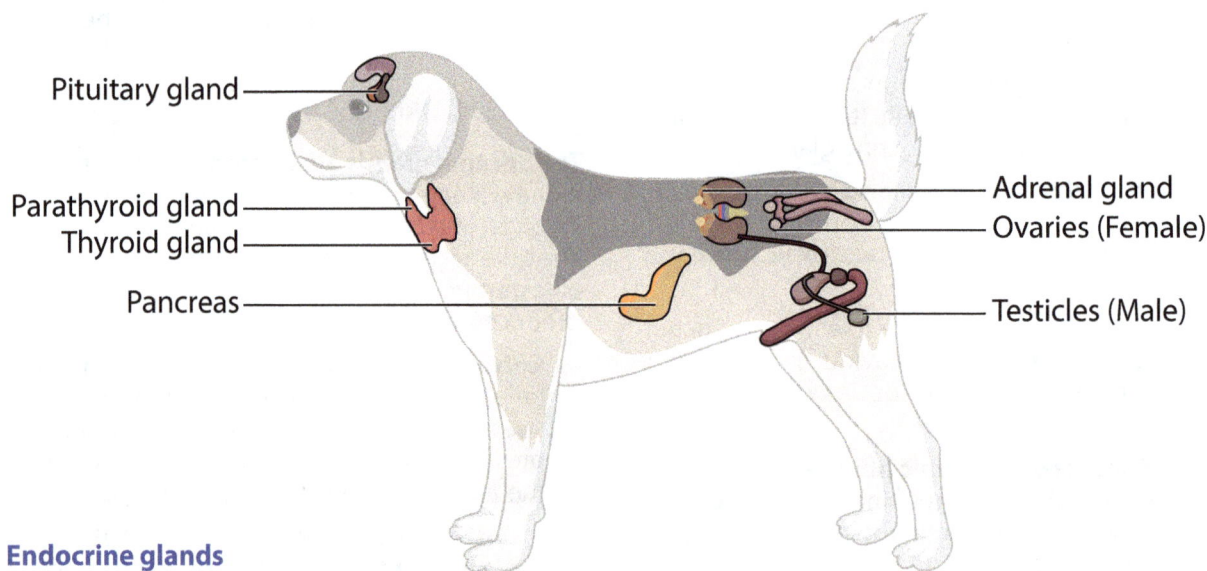

Pituitary gland

Parathyroid gland
Thyroid gland

Pancreas

Adrenal gland
Ovaries (Female)

Testicles (Male)

Endocrine glands

Hormonal mechanisms

Hormones tend to fall into one of two categories:

- **Circulating hormones** travel through the bloodstream to target areas around the body. A good example of this is insulin and glucagon. These hormones are produced in the pancreas but are required in cells around the body.
- **Locally acting hormones** are released and act in the area of the body where they are needed. An example of this is **histamine** which is produced by the immune system in response to injury or allergens. Histamine triggers an inflammation response, increasing blood flow to the area.

In both cases, hormones need specific receptors on target cells to trigger a response. When a hormone makes contact with one of these receptors, the target cell proceeds to carry out the instruction. For example, when insulin binds to insulin receptors on cells, channels open allowing glucose into the cell where it is used to produce ATP.

Homeostasis

Homeostasis is the process by which conditions in the body are kept within an acceptable range. There are a number of different homeostasis mechanisms.

Glucoregulation

Glucoregulation is the control of blood sugar levels in an animal's body. Concentrations of glucose that are too high or too low can be very harmful to the animal's health. When glucoreceptors in the pancreas detect a change in blood glucose levels, specialised cells called the **islets of Langerhans** produce

- **insulin** to **lower** blood sugar levels

- **glucagon** to **increase** blood sugar levels.

Glucoregulation is an example of a negative feedback homeostatic function.

Osmoregulation

Water is essential for many processes in the animal body. The hormone **antidiuretic hormone** (**ADH**) controls how much water is released in urine by the kidneys.

- When the body is dehydrated, the pituitary gland releases more ADH reducing the amount of water that leaves the body. In this case, the body produces small amounts of very concentrated urine.

- When the body has enough water less ADH is produced, increasing the amount of water that leaves the body. This produces much larger amounts of very dilute urine.

Thermoregulation

The animal body functions best when it is within an ideal temperature range. The ideal range is determined by factors such as species, metabolic rate and environment. The majority of mammals have a body temperature around 38°C whilst birds have a higher body temperature around 40°C.

Animals have evolved various mechanisms to control their body temperature:

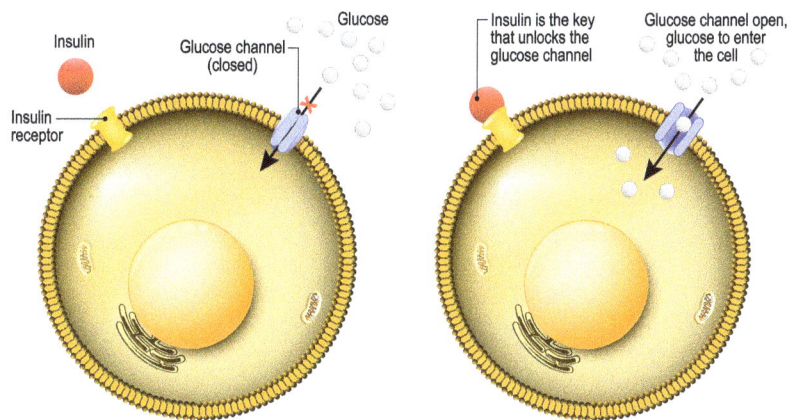

How insulin works

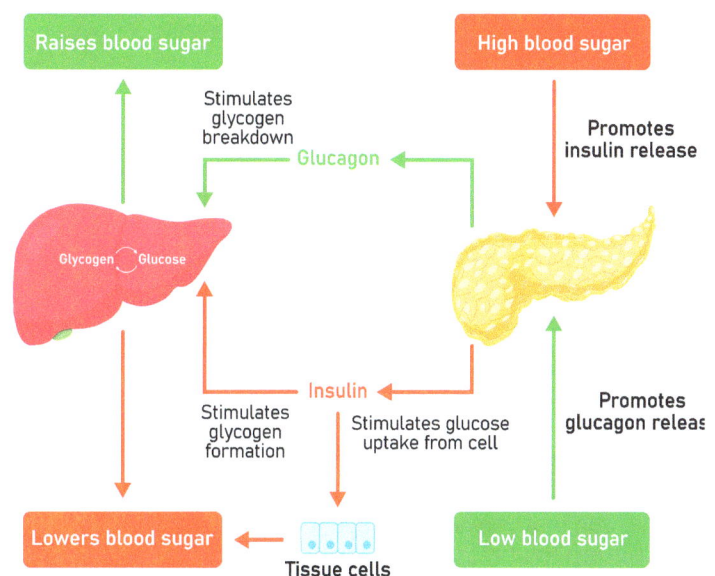

Glucoregulation

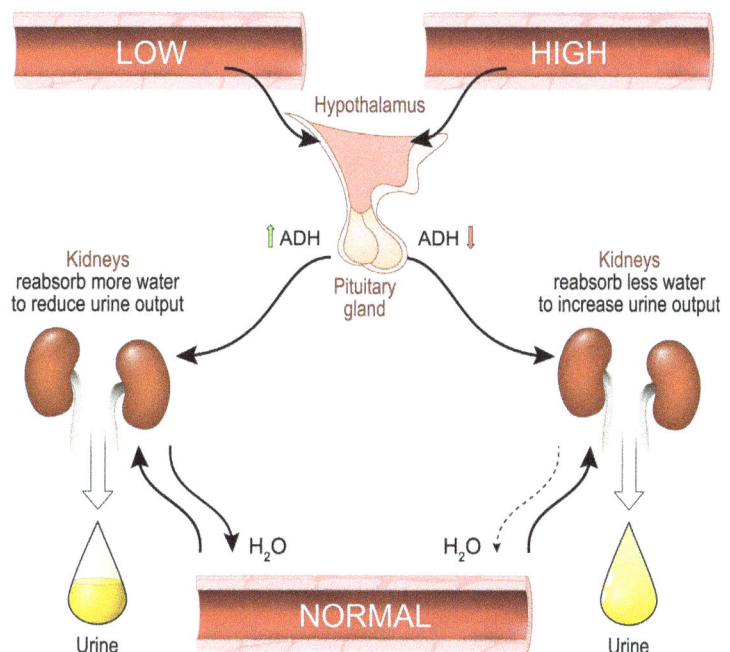

Osmoregulation

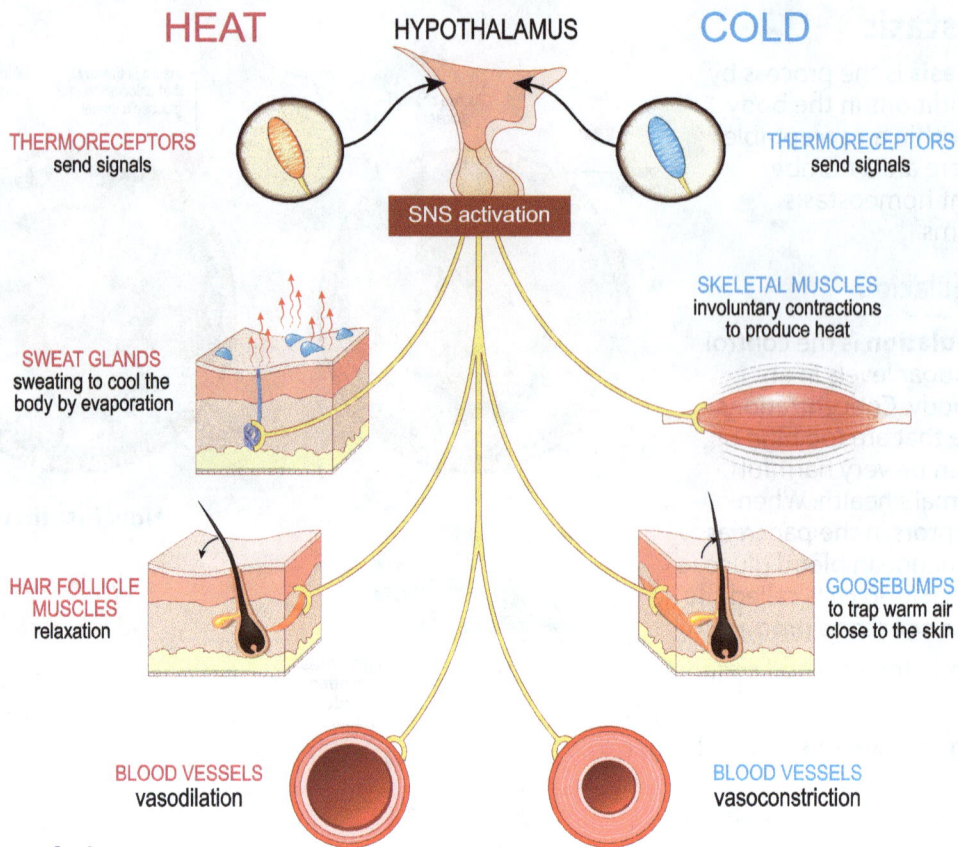

HEAT   HYPOTHALAMUS   COLD

THERMORECEPTORS
send signals

THERMORECEPTORS
send signals

SNS activation

SKELETAL MUSCLES
involuntary contractions
to produce heat

SWEAT GLANDS
sweating to cool the
body by evaporation

HAIR FOLLICLE
MUSCLES
relaxation

GOOSEBUMPS
to trap warm air
close to the skin

BLOOD VESSELS
vasodilation

BLOOD VESSELS
vasoconstriction

Thermoregulation

Endothermic species generate and regulate their body heat through metabolic processes such as shivering, sweating, vasodilation and vasoconstriction. Adaptations such as fur, feathers and fat layers help to trap heat close to the body.

Ectothermic species rely on external sources of heat such as basking on a warm rock or hiding in shade to maintain their body temperature.

Thermoregulation is vital in animals as many physiological processes rely on the body being within a certain temperature range.

> ### Important terms!
>
> **ATP**: Adenosine triphosphate, the energy currency of the cell.
>
> **Negative feedback**: A homeostatic function that brings the body back to normal levels. Example: glucoregulation.
>
> **Endothermic**: Species that produce and maintain their own body heat.
>
> **Vasodilation**: Blood vessels expand to allow more blood to flow through them.
>
> **Vasoconstriction**: Blood vessels contract to reduce the flow of blood through them.
>
> **Ectothermic**: Species that rely on external sources of heat to maintain their body temperature.

Recap Questions

1. What is the primary function of the endocrine system?

2. Name five endocrine glands and their location the animal body. Explain their function.

3. Which gland controls the endocrine system?

4. What are hormones?

5. How do circulating hormones differ from locally acting hormones?

6. What is homeostasis?

7. Explain how endothermic animals maintain their body temperature.

8. Why is thermoregulation important?

9. How do animals control the amount of water in the body?

6.5 The structure and function of the musculoskeletal system in relation to animal physiology

The **musculoskeletal system** is essential in the animal body for:

- providing support
- protection of the organs
- locomotion (movement)
- haematopoiesis (blood cell production)
- mineral storage.

It consists of bones, cartilage, muscles, tendons and ligaments which all work together to allow animals to maintain posture and move in order to interact with their environment. In vertebrates, the musculoskeletal system has huge variety based on how animals have adapted to live.

The skeleton

In the majority of animal species, the skeleton is made of calcified bone, which contains the following:

- The exterior of the bone is made up of hard **compact bone**, which provides rigidity and strength.
- The interior contains **spongy bone**, which provides structural support whilst reducing weight.

- **Bone marrow** stores fat, acts as an energy reserve and produces blood cells.
- **Blood vessels** supply nutrients and oxygen, whilst **nerves** provide sensory information related to pain and injury.

The **epiphyseal line** is the remains of the growth plate. It is found only in long bones and marks the point where bone growth took place in juvenile animals.

Bone cells

Osteoblasts are bone-forming cells that secrete a matrix which hardens into a calcified tissue.

Osteocytes are mature bone cells.

Osteoclasts are cells that break down bone in order to absorb the minerals stored there for use elsewhere in the body.

Skull

The **skull** protects some of the most delicate organs in the animal body: the brain and eyes. It is comprised of three parts: The **cranium**, the **mandible** and the **maxilla**.

Vertebrae

The **spine** is comprised of bone discs called **vertebrae**, separated by discs of cartilage. The spine is very flexible allowing animals to bend and twist. The number of vertebrae differs between species and how they have adapted to live. The spine is separated into five distinct sections.

Cervical

The **cervical spine** (also known as the neck) supports the head of the animal.

- The **first vertebrae** in the cervical spine is called the **atlas**. This supports the skull and allows a nodding motion.
- The **second vertebrae** is called this **axis** and this allows rotational movement.

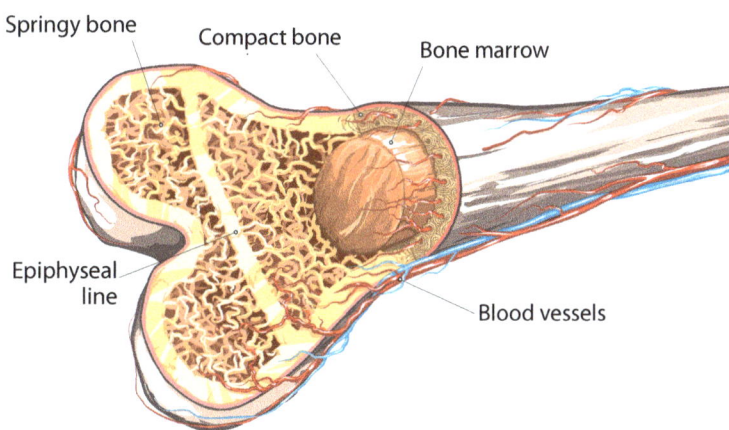

Springy bone · Compact bone · Bone marrow · Epiphyseal line · Blood vessels

The structure of bones

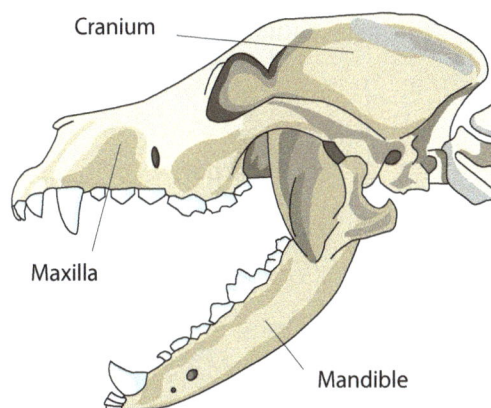

Cranium · Maxilla · Mandible

The skull

The majority of mammals have **seven cervical vertebrae**. The exceptions to this rule are manatees (9) and sloths (5 or 6).

Thoracic

The **thoracic spine** is attached to the rib cage. This protects the heart and lungs. Movement is limited in this area of the spine.

Lumbar

The **lumbar spine** is the vertebrae that make up the lower back. These are short, thick vertebrae adapted to provide strength to support the body.

Sacral

The **sacral spine** is the pelvic region and consists of five vertebrae fused into one bone. This gives strength to the pelvis and provides attachments points for pelvic muscles.

Coccygeal

The **coccygeal spine** is the vertebrae that make up the tail. There is huge variation in the number of vertebrae between different species in this area of the spine.

Dog skeleton

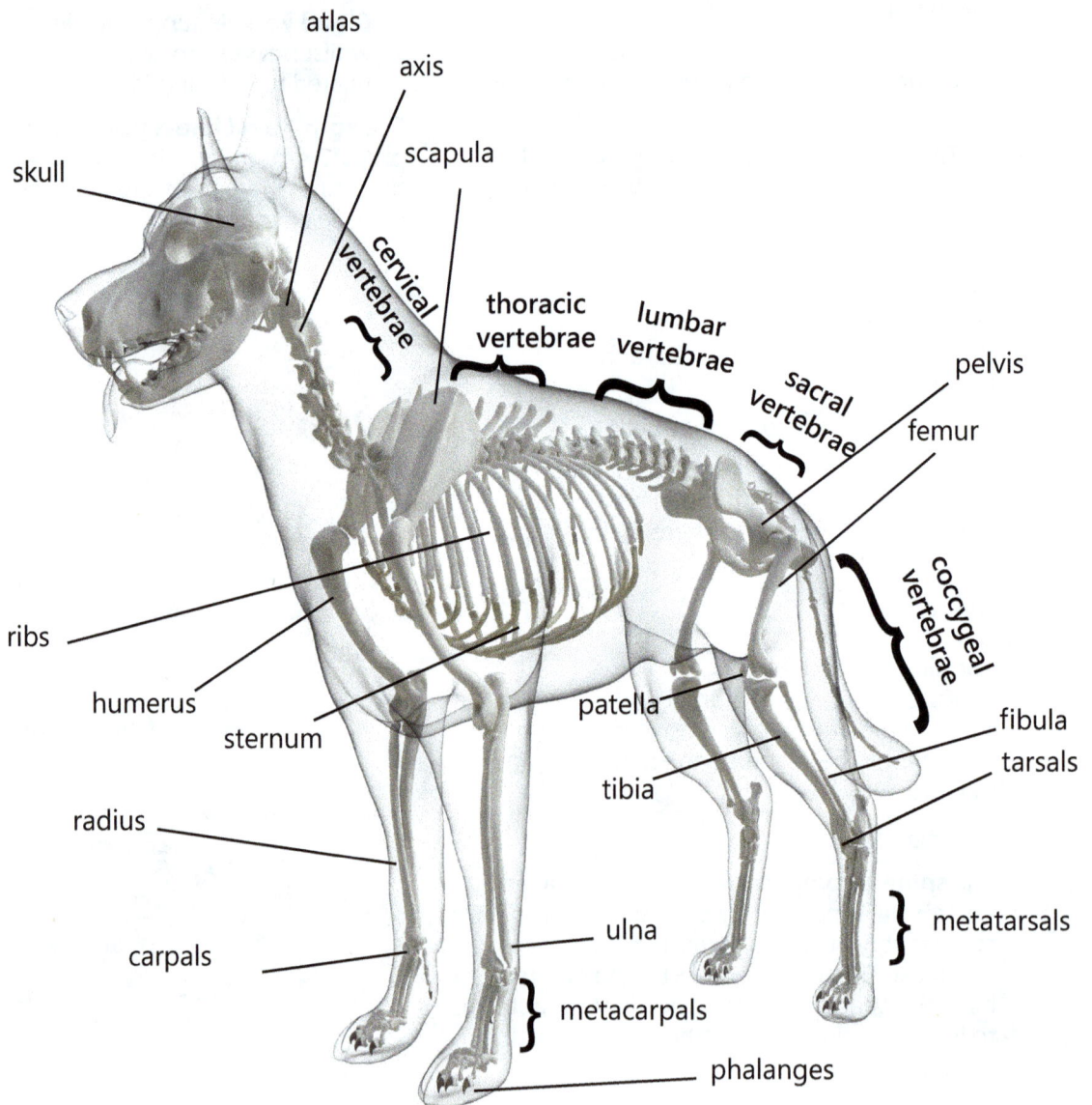

Skeletal adaptations in mammals

Cats have a bone in their throat called the **hyoid** bone. This is linked to sound production in cats.

The **baculum** is a bone found in the penis of many mammals. This bone provides support to the penis and assists the animal with mating. It is found in many animal species where the penis is very large in relation to body size. Without a baculum, these species would struggle to maintain erection due to the amount of blood required.

Aquatic mammals

Cetaceans (whales, dolphins and porpoises) evolved from mammals who lived on land. They have several adaptations to living in the water, including:

- Lighter skeletal bones which are not as strong, because they do not need to support the full mass of the animal

- Much longer phalanges in the forelimbs, so their flippers can act like paddles

- They have lost their hindlimbs because they use their tail provide power, and are more streamlined without them.

- Neck vertebrae are fused together, to withstand the forces on the neck from diving.

In whales, the ribs are unattached to the sternum, to allow them to collapse their lung when diving.

Flying mammals

Bats are mammals that can fly. Flying squirrels glide rather than really fly. Skeletal adaptations include:

- Thinner, lighter bones that are denser than similar-sized mammals. This allows the animal to withstand the forces of flying.

- Bats have very long metacarpals and longer phalanges on the forelimbs, which are attached to their wings in order to extend them and allow them to fly.

- Bats also have an extended ridge of bone on the sternum, called a **keel**, to which the muscles responsible for flying are attached.

A bat in flight – notice their long forelimbs

Hopping mammals

Hopping allows animals to move rapidly and quickly change direction, and is thought to have evolved to allow small mammals to avoid predators. The most common hopping mammal is the rabbit, whose adaptations include:

- The tibia and fibula in the hindlimbs are relatively much longer. The forelimbs are much smaller than the hindlimbs

- There is greater separation between the lower vertebrae, as well as an extended lumbar, to allow for more flexibility and extension when hopping. This gives the spine a curved shape but the drawback is that this curved spine is prone to breaking.

A rabbit's hindlimbs are adapted for hopping

Running animals

Running animals, such as horses, lions and cheetahs, are more formally known as cursorial animals. Running mammals tend to be relatively large.

Cheetahs have evolved several skeletal adaptations which allow them to run very quickly.

- A small skull, in order to be streamlined.

- A very flexible spine and elongated legs mean the cheetah is able to curl up tightly before bounding forward. This gives both speed and a very long stride, allowing the cheetah to reach speeds of up to 70mph (120km/h).

Horses are adapted for long distance running:

- They have fused their ulna and radius, and a fused tibia and fibula, so their limbs can take the impact of running.
- Their limbs are relatively longer, so they can cover more ground with each stride.
- They have a large rib cage to house larger lungs, which are needed for high-intensity activity like running.

Predator and prey

Big cats

Big cats, such as lions or tigers, have evolved a number of adaptations.

- A flexible lumbar spine, allowing them to stretch and contract in a bounding motion. This enables short bursts of speed when ambushing prey.
- Their scapula (shoulder blade) is loosely attached with muscles, increasing limb mobility.
- Strong, robust limb bones reduce the likelihood of breaking during pounding and grappling prey.
- Cats have retractable claws. Bones in the toes and digits allow claws to be sheathed when not in use, to prevent wear or damage. Claws are extended using flexor tendons.
- A shortened rostrum (snout) allows for a powerful bite.
- The shape of the skull (large sagittal crest and zygomatic arches) provides attachments for strong jaw muscles, increasing bite strength.
- A long, flexible tail acts as a counterbalance during high-speed chases, enabling stability and manoeuvrability.

Ungulates

Ungulates such as horses, deer or antelope have evolved adaptations to escape predators through speed and endurance.

- Long metacarpals and metatarsals give elongated limbs, which increase stride length for efficient running.
- Many ungulates have evolved a single toe (such as horses), whilst others have two toes (deer and antelope).
- The radius and ulna have fused together

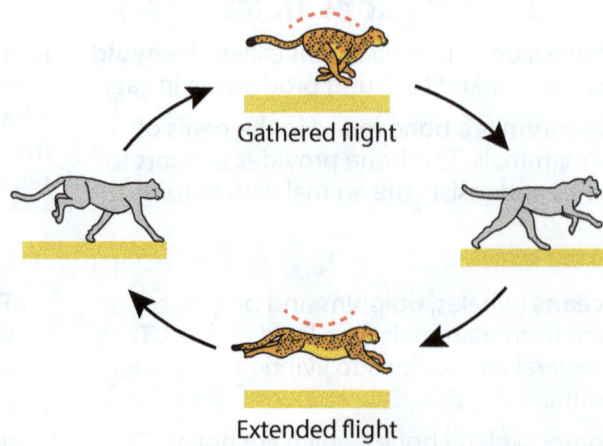

Gathered flight

Extended flight

The running action of a cheetah

A female lion hunting

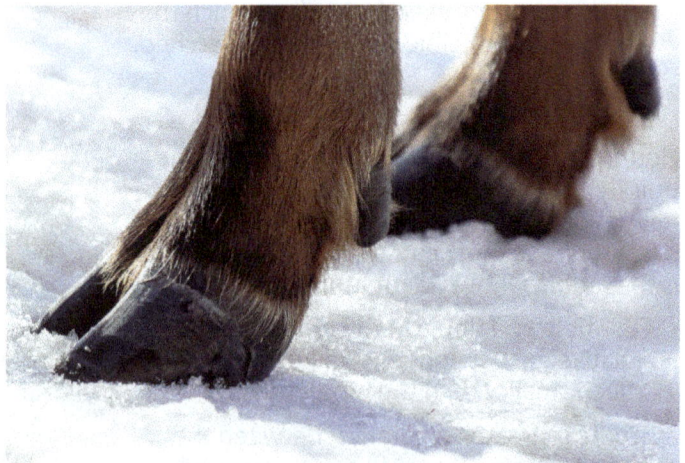

Deer have evolved to walk and run on two toes

into the **radioulnar**, whilst the tibia and fibula have fused into the **tibiofibular**. This produces larger, stronger bones which are less likely to be damaged during sustained running.

- Horses have locking joints which means they can sleep standing up.

Non-mammalian adaptations

Aves

Birds have **pneumatic bones**.

- These are bones with a hollow, honeycomb structure which links closely to their respiratory system.

- The structures that provide support inside these bones are called trabeculae.

- These bones do not have bone marrow.

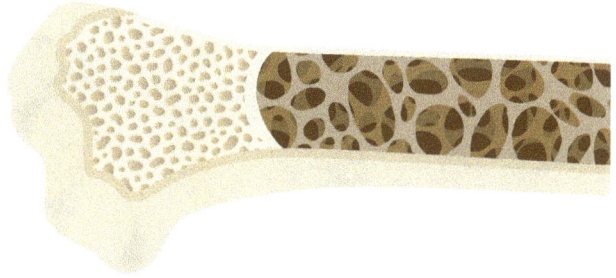

Pneumatic bone in a bird

Fish

Sharks and rays have a skeleton made of **cartilage** which is lighter and more flexible than bone. This often means that these fish can swim faster than bony fish of a similar size.

Reptilia

Reptiles have many skeletal adaptations.

- Snakes have evolved many small vertebrae without limbs. This means they are very flexible. Snakes have anywhere between 150 and 450 vertebrae depending on the species.

- Tortoises have evolved a hardened exterior shell to give them protection. It is made from two parts:

  » The top part of the shell (the dorsal shell) is called the carapace. It is made from bones that have fused together, including the ribs and spine. It is covered in a keratin layer of large scales called **scutes**.

  » The underside of the shell is the **plastron**. This is made from the bones of the fused pectoral girdle.

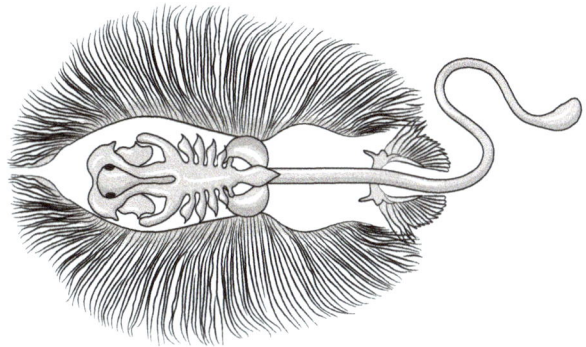

Cartilage skeleton of a ray

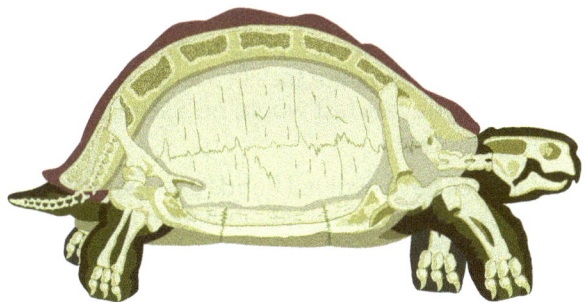

A tortoise skeleton

Invertebrates are very different to vertebrate species in that they do not have an internal skeleton. Instead, they have an exoskeleton which gives the body shape and protection. This exoskeleton is made from chitin (insects and arachnids) or calcium carbonate (molluscs, crabs and lobsters).

Pentadactyl limb

In vertebrate species, the pentadactyl limb (meaning five digits), is an example of a homologous structure. The similarities of these limbs between species tell us that this has evolved from a common ancestor but it has adapted to perform specific functions:

- Aquatic species such as whales and dolphins have evolved forelimbs modified into flippers for swimming.

- Flighted species such as bats have evolved elongated forelimbs that form wings. The length of the bones in wings support the body in flight.

- Species such as rabbits and kangaroos have adapted and elongated hindlimbs, which allow for powerful propulsion.

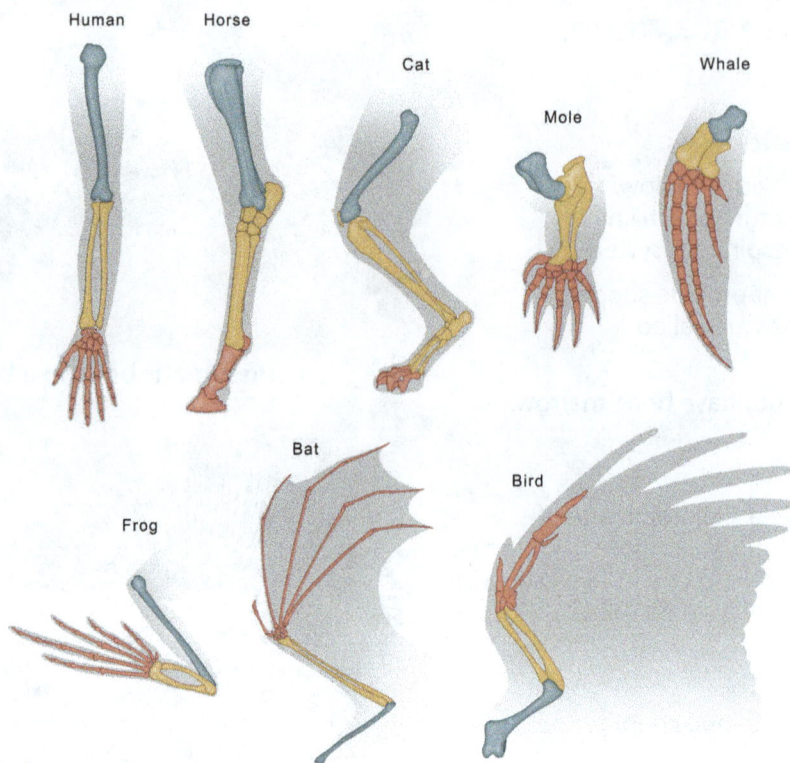

Pentadactyl limbs

- Horses have the same bones as other vertebrates in their legs however the bones have fused into a single, large hoof. This gives strength to the limb and allows the animal to run at speed. This is essential for horses as they are such large, heavy animals. Having five separate toes in a large species that moves very fast would be a severe disadvantage.

- Primates have evolved hands and feet that enable climbing and manipulation of objects. This is essential for tool use and interacting with the surrounding environment.

> **Important terms!**
>
> Calcified: A substance that has been hardened by calcium.
>
> Haematopoiesis: The production of blood cells.
>
> Cursors: An animal that is adapted for running.

Movement

Movement is achieved when muscles pull against bones of the skeleton. Muscles are **contractile tissue** meaning they can only pull, not push. To enable a limb to move in two directions, skeletal muscle works in pairs:

- As muscle 1 contracts, muscle 2 relaxes. This moves a limb in one direction.

- To move the limb in the opposite direction, muscle 2 contracts and muscle 1 relaxes.

Muscles are attached to bones via **tendons**. These are incredibly strong, rope-like fibres made of connective tissue.

To help support the joint, **ligaments** connect bone to bone.

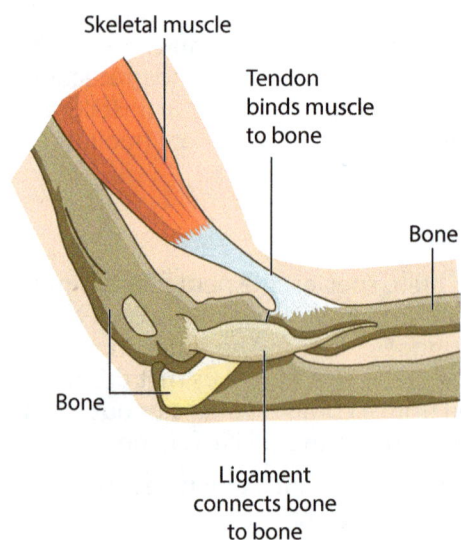

Skeletal muscle

Tendon binds muscle to bone

Bone

Bone

Ligament connects bone to bone

Muscles, tendons and ligaments

Joints

Joints are found throughout the animal body.

- **Fibrous joints** are permanent connections where bones have fused together (e.g. the skull).

- **Cartilaginous joints** have areas of cartilage separating bones (e.g. the spine).

- **Synovial joints** allow movement. These joints have a cavity between the bones filled with **synovial fluid** which acts as a lubricant.

There are several types of synovial joints:

Joint	Structure	Movement	Examples
Ball and Socket		Flexion and Extension Abduction and adduction Rotation Circumduction	Hip and shoulder
Hinge		Flexion and extension	Elbow and knee
Pivot		Rotation	Axis (neck)
Condyloid		Flexion and extension Abduction and adduction Circumduction	Wrist
Saddle		Flexion and extension Abduction and adduction Circumduction	Thumb
Gliding		Gliding	Wrist

Important terms!

Flexion: Decreasing the angle between two bones.

Extension: Increasing the angle between two bones.

Abduction: Moving a limb away from the midline of the body.

Adduction: Moving a limb toward the midline of the body.

Rotation: Turning a bone around its axis (like turning a screwdriver).

Circumduction: A circular motion combining flexion, extension, abduction and adduction.

Gliding: Sliding motion between two flat bones.

Circumduction

1. State three variations of the pentadactyl limb and explain how each determines the type of movement the animal can perform.

2. What is the function of the hyoid bone in cats?

3. Explain three adaptations that allow big cats to hunt.

4. What is the difference between tendons and ligaments?

5. Give five functions of the skeleton.

6. Name the sections of the spine.

7. Identify three synovial joints, giving an example of where they are found in the body.

8. What is the function of the baculum? Why do only certain species have one?

6.6 The structure and function of the nervous system in relation to animal physiology

The **nervous system** is a complex network that coordinates all body systems, integrates sensory input and controls responses. The vertebrate nervous system comprises the **central nervous system** (**CNS**) and the **peripheral nervous system** (**PNS**) which connect the brain to every part of the body. Different parts of the brain regulate specific functions and enable the animal to interact with its environment through specially adapted senses.

The structure and function of the brain

The brain is split into two halves called **hemispheres**. Different sections of the brain manage different functions and senses.

Forebrain

Thalamus: Acts as a relay centre, processing and transmitting sensory and motor signals to the cerebral cortex.

Hypothalamus: Regulates essential functions such as body temperature, hunger, thirst, sleep, and hormone release by interacting with the endocrine system.

Cerebral Cortex: The largest part of the brain responsible for higher-order functions like thought, memory, decision-making, voluntary movements and processing sensory input.

Limbic System: Includes structures like the amygdala and hippocampus, involved in emotions, memory and learning.

Midbrain

Reticular Formation: A network of neurons important for regulating sleep-wake cycles, arousal and attention.

Neuron Receptors: Detect changes in the environment and relay information, playing a key role in sensory and motor signal integration.

Hindbrain

Medulla Oblongata: Controls vital involuntary functions such as breathing, heart rate and blood pressure.

Cerebellum: Coordinates fine motor skills, balance and posture.

Pons: Bridges information between the brain and spinal cord; also involved in breathing and facial expressions.

The nervous system

Central Nervous System

The **central nervous system** (**CNS**) is comprised of the brain and spinal cord. It interprets sensory input and formulates responses. The spinal cord acts as a communication highway between the brain and peripheral nerves.

Peripheral Nervous System

The **peripheral nervous system** (**PNS**) connects the central nervous system to the rest of the body, allowing the brain to coordinate each body system so they can all work together.

- Afferent (sensory) neurons carry information from sensory organs to the central nervous system. These nerves collect information from the external environment to be processed by the central nervous system.

- Efferent (motor) neurons transmit commands from the central nervous system to muscles and glands, allowing the body to react to external stimuli and perform voluntary movements.

Autonomic Nervous System

This system controls involuntary functions. It is divided into two main sections:

- **Sympathetic System**: Activates the "fight or flight" response during stressful situations. This includes, increasing the heart and breathing rate, dilating pupils and inhibiting digestion. This system is vital for survival in situations where the animal's life is threatened.

- **Parasympathetic System**: Promotes "rest and digest" functions, for example slowing the heart and breathing rate, relaxing the pupils and enhancing digestion. This allows the animal to conserve energy and digest food.

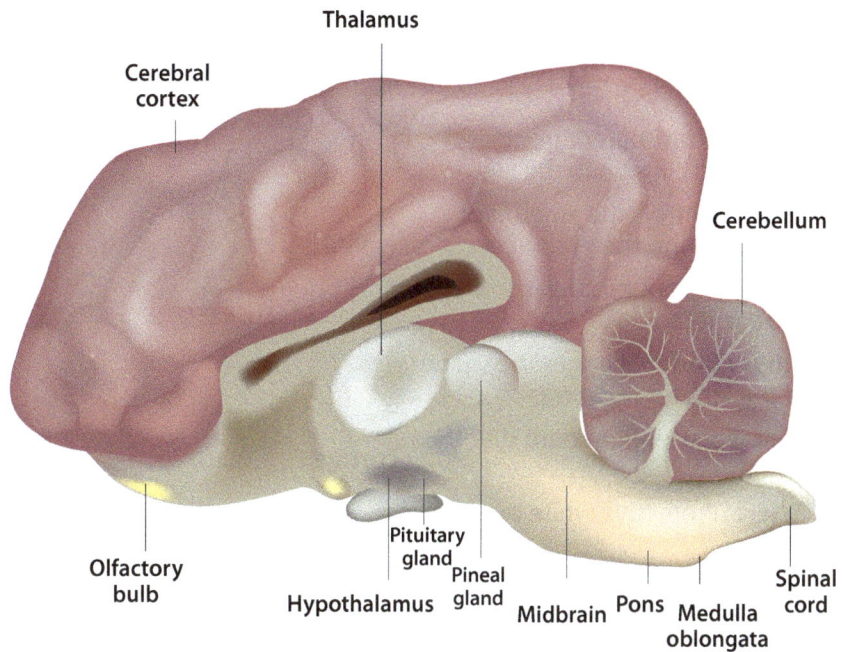

The structure of the canine brain

Cerebral cortex — Thalamus — Cerebellum — Olfactory bulb — Hypothalamus — Pituitary gland — Pineal gland — Midbrain — Pons — Medulla oblongata — Spinal cord

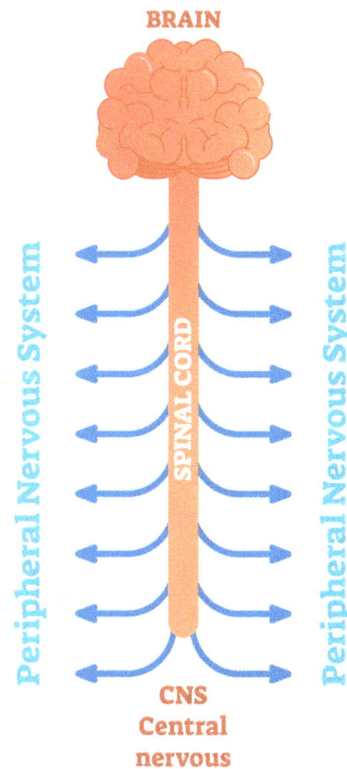

BRAIN

Peripheral Nervous System

SPINAL CORD

Peripheral Nervous System

CNS Central nervous

The CNS and PNS

Important terms!

Afferent neurons: Sensory neurons that carry information from the external environment to the central nervous system.

Efferent neurons: Motor neurons that carry instructions from the central nervous system to muscles and glands.

Senses

Eyes

Cornea: Transparent front layer that bends light into the eye.

Pupil: Controls the amount of light entering the eye.

Iris: Adjusts pupil size and gives the eye its colour.

Ciliary Body: Changes the shape of the lens for focusing.

Lens: Focuses light onto the retina.

Sclera: Protective white outer layer of the eye.

Retina: Contains **rod cells** (to detect low levels of light) and **cone cells** (to detect colour).

Choroid: Supplies blood to the retina.

Fovea: A region of the retina with a high concentration of cone cells for sharp vision.

Optic Disc: The "blind spot" where the optic nerve exits the eye.

Optic Nerve: Transmits visual information to the brain.

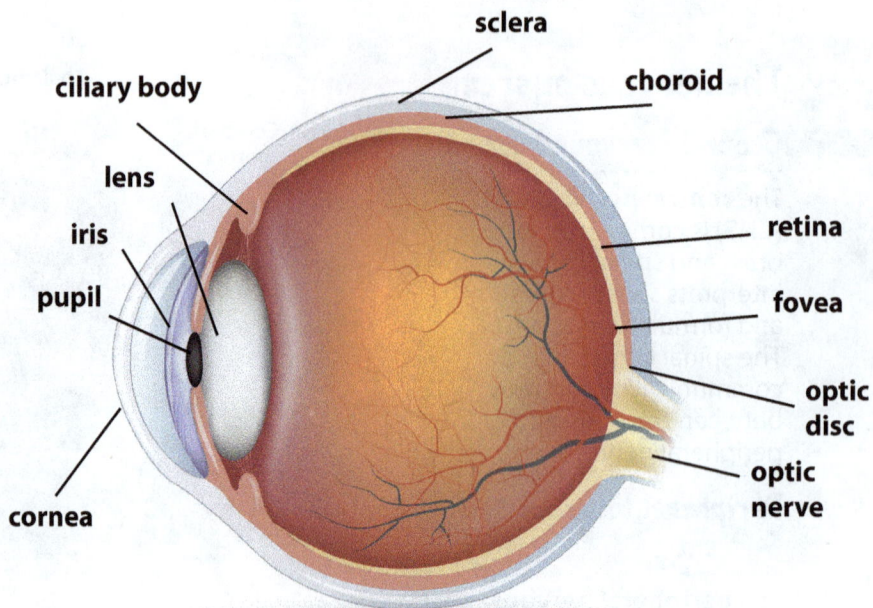

The structure of the eye

Rectus Muscles: Control eye movement.

Tapetum lucidum: A reflective layer behind the retina that reflects light back into the retina, effectively doubling the amount of light available. This is an adaptation for nocturnal species.

Ears

Pinna: The outer part of the ear that collects and directs sound waves into the ear.

Auditory Meatus: The ear canal, amplifying sound.

Tympanic Membrane: Vibrates in response to sound waves.

Ossicles (Malleus, Incus, Stapes): Tiny bones found in the inner ear that amplify sound vibrations and transfer them to the cochlea.

Oval and Round Windows: Transmit vibrations into the cochlear fluid.

Cochlea: Converts sound waves into nerve impulses.

Organ of Corti: Detects sound vibrations and sends signals to the brain via the **Cochlear Nerve**.

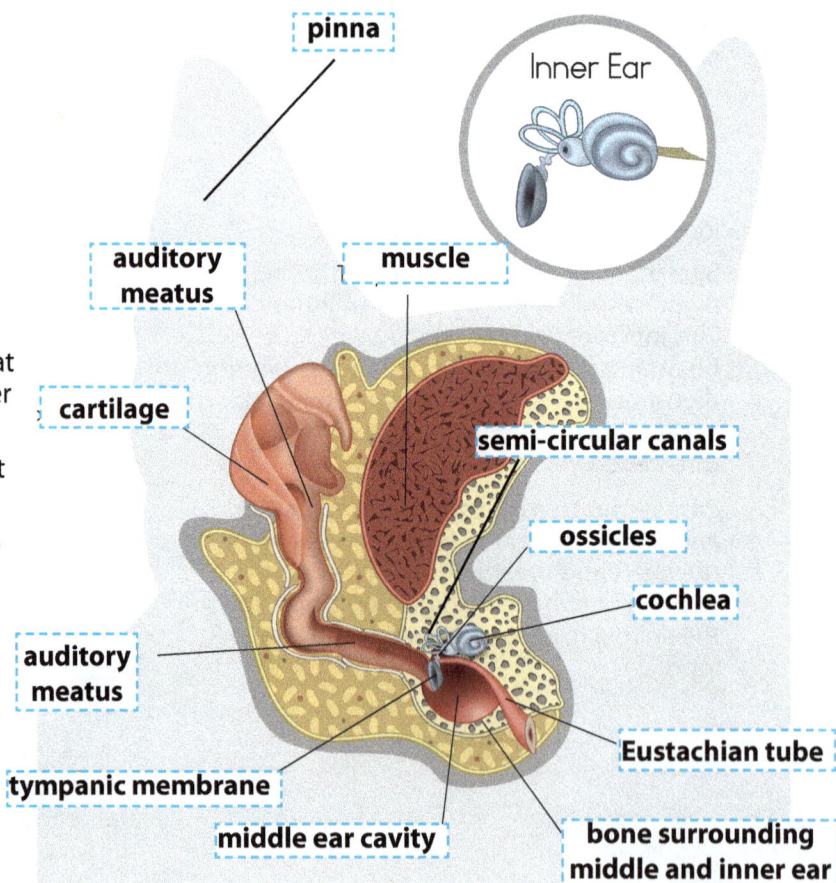

The structure of the ear

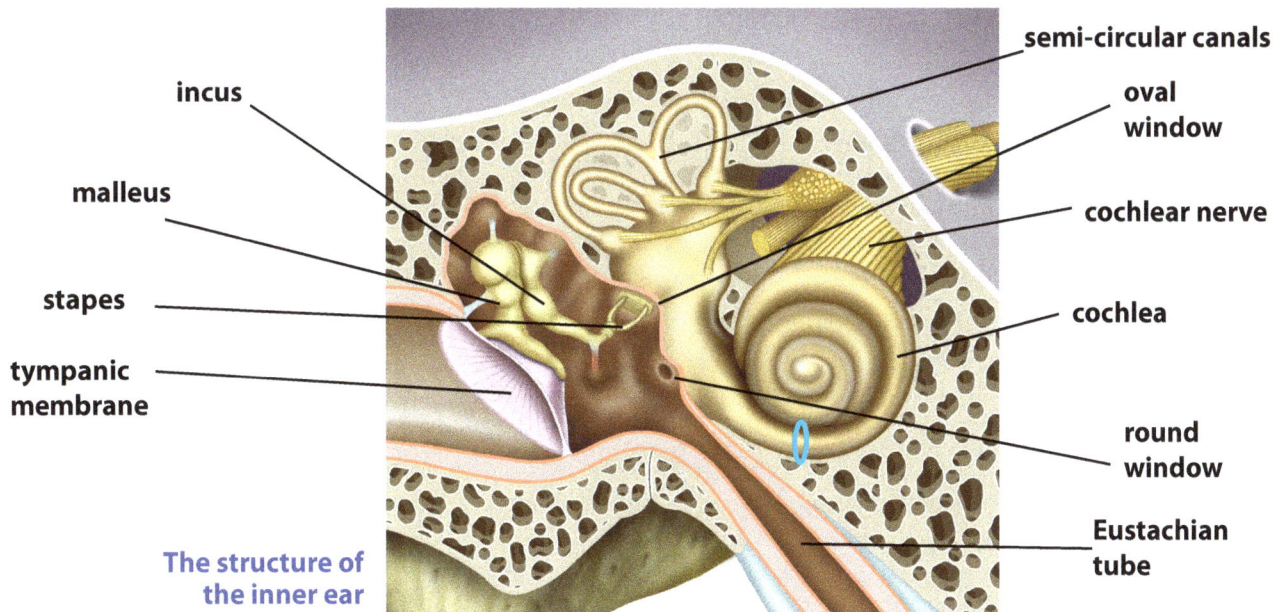

The structure of
the inner ear

Labels for inner ear diagram: incus, malleus, stapes, tympanic membrane, semi-circular canals, oval window, cochlear nerve, cochlea, round window, Eustachian tube

Nose

Nasal Chambers: The space from the nostrils to the back of the throat

Turbinates: These structures are made of bones, tissues, blood vessels and olfactory receptors. They filter and warm the air, increasing the contact of odour molecules with the olfactory receptors.

Olfactory Nerve: This sends messages from the olfactory receptors to the olfactory bulb.

Olfactory Bulb: Processes the information about chemical molecules in the air that was detected by the olfactory receptors. This information is sent on to the brain, triggering the sense of smell.

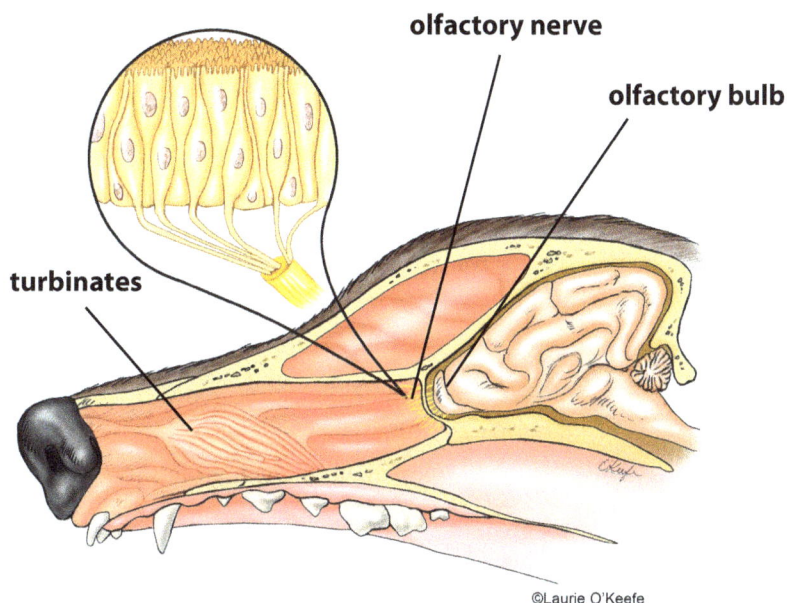

Labels for nose diagram: olfactory nerve, olfactory bulb, turbinates

©Laurie O'Keefe

The structure of the nose
© Laurie O'Keefe, reproduced with kind permission

Mouth

Taste Buds: Contain receptors for detecting sweet, sour, salty, bitter, and umami (savoury) flavours.

Soft and Hard Palates: Aid in swallowing, and separates the oral and nasal cavities.

Touch

Skin Receptors: Detect pressure, temperature, pain and vibration, providing tactile feedback.

The left-hand image of a tiger shows the hard palate (towards the very front of the mouth) and soft palate. The right-hand image of a dog shows taste buds on the tongue, visible as small indentations on the surface.

Sensory adaptations

Predator vs prey

Predators often have forward-facing eyes for depth perception (**bifocal vision**) and enhanced hunting ability whilst prey species typically have side-facing eyes (**monofocal vision**) for a wider field of view to detect threats.

Tactile organs

The duck-billed platypus has a bill with tens of thousands of mechanoreceptors on it. The mechanoreceptors allow the animal to detect obstacles, feel its way and find prey.

Taste and smell

Jacobson's organ (vomeronasal organ) detects pheromones, aiding in reproduction and social behaviours. Some species of reptile such as snakes have a highly evolved Jacobson's organ which they use to locate prey.

Electroreception

Many species use sensors to detect electrical signals in the world around them.

- Sharks' snouts are covered in small pits called **Ampullae of Lorenzini**. These are able to detect electrical fields generated by muscle movement in prey.

- Ampullae of Lorenzini are also found in other cartilaginous fish.

- Platypus beaks are covered in approximately 100,000 sensors that detect electrical signals from prey.

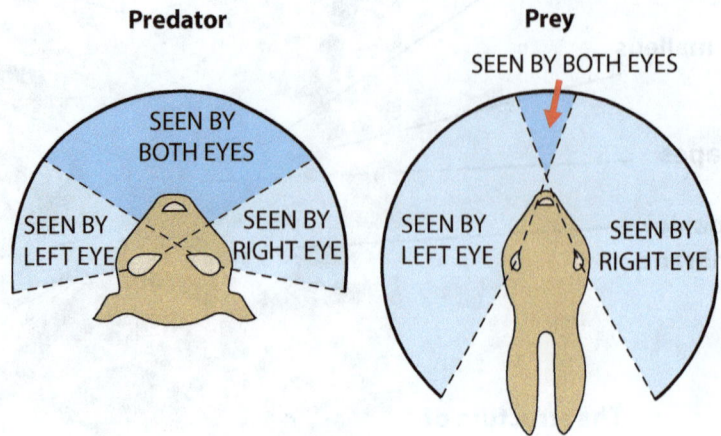

Predator

SEEN BY BOTH EYES

SEEN BY LEFT EYE

SEEN BY RIGHT EYE

Prey

SEEN BY BOTH EYES

SEEN BY LEFT EYE

SEEN BY RIGHT EYE

Eye position in predatory and prey species

Duck-billed platypus

Brain

Nerve

Jacobson's organ

Tongue

Jacobson's organ in a snake

Shark Ampullae of Lorenzini

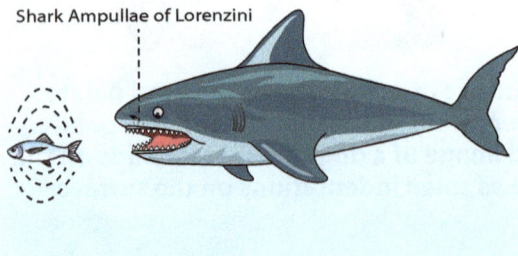

Ampullae of Lorenzini on a shark

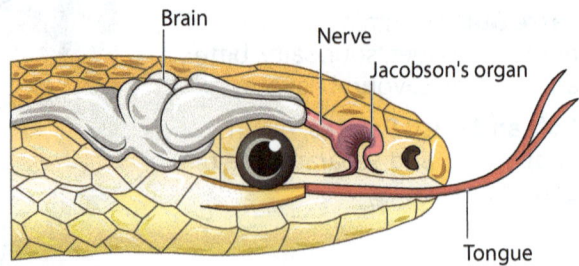

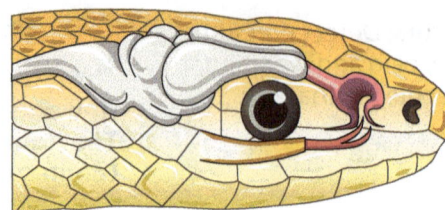

Echolocation
Bats Navigate and Identify

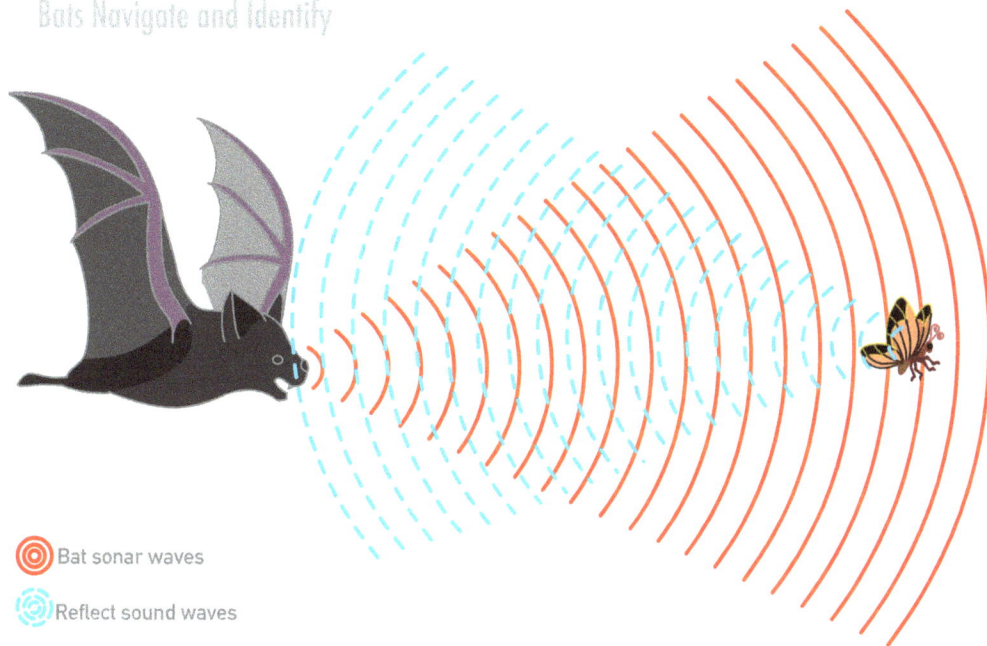

Bat sonar waves

Reflect sound waves

Echolocation in bats

Echolocation

Bats and dolphins emit high frequency sound waves and analyse the echoes from these to navigate and hunt. This is especially useful in dark or murky environments where vision is limited.

Lateral line

A line of pores found down the side of fish which detects water currents and vibrations.

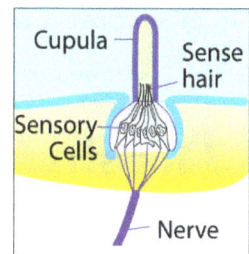

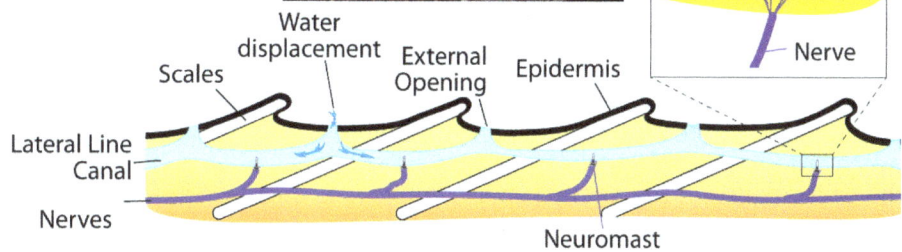

Vibrissae

Vibrissae (or whiskers) are specialised sensory hairs which are used for sensing touch and movement. Animals such as seals use them to detect prey and changes in water currents under the water whilst animals such as cats and rats use them to feel the environment around them detecting space and movement.

Scales
Water displacement
External Opening
Epidermis
Lateral Line Canal
Nerves
Neuromast
Cupula
Sense hair
Sensory Cells
Nerve

Lateral line in fish

A seal's vibrissae

6.7 The structure and function of the excretory system in relation to animal physiology

The excretory system plays a vital role in maintaining homeostasis by removing metabolic waste products, excess water, and toxins from the body. It ensures the internal environment remains stable by regulating essential factors such as fluid balance, electrolyte levels and blood pH.

The kidneys are the primary organs of this system, working in coordination with the bladder, ureters and urethra to filter blood and excrete urine.

Beyond waste removal, the excretory system interacts with other systems, such as the endocrine and digestive systems, to manage hormone levels and process metabolic byproducts. This system also has fascinating adaptations in different species, highlighting its evolutionary significance in various environments.

Structure and function of the excretory system

Kidneys

The **kidneys** are the primary organs of the excretory system. Their role is to filter blood in order to remove waste products, excess water and toxins.

The kidneys help the body to maintain homeostasis by regulating electrolytes (salts), blood pressure and pH levels.

Within the kidney are millions of filtration units called **nephrons**. Each nephron has the structure shown in the diagram.

- Blood pressure forces plasma, small molecules (such as salts) and waste products (such as urea) through the **glomerulus** into the **Bowman's capsule**.

- The glomerulus acts as a sieve, allowing these small particles to pass through whilst stopping larger molecules such as proteins and blood cells. These small particles are removed from the bloodstream. This process is called **ultrafiltration**.

- As the fluid passes through the nephron, essential substances like water, glucose, amino acids and ions are reabsorbed into the bloodstream. This process is called **reabsorption**. The amount of water and substances that are reabsorbed is determined by the body's current needs and controlled by hormones.

Antidiuretic hormone (**ADH**) is produced by the **hypothalamus** and released by the **posterior pituitary gland**. It increases the amount of water reabsorbed by the body. Higher ADH levels occur if the animal is dehydrated.

Aldosterone is secreted by the **adrenal glands** and increases the amount of sodium ions (Na^+) being reabsorbed. This influences blood pressure and fluid balance.

Parathyroid hormone is released when calcium levels are low and causes calcium ions (Ca^{2+}) to be reabsorbed.

Ureter

The **ureter** is a tube that transports urine from the kidneys to the bladder via peristalsis.

Bladder

The **bladder** is a highly elastic storage organ for urine. The bladder has **two sets of sphincter muscles** that control how and when urine is released:

- **External sphincter muscles** are skeletal muscle under voluntary control. The animal must relax these muscles to release urine.

- **Internal sphincter muscles** are regulated by the autonomic nervous so are involuntarily controlled. These are made of smooth muscle and relax when the bladder is sufficiently filled.

Urethra

The urethra is a tube through which urine exits the body. It is much longer in males as it passes through the length of the penis. In males, sperm also exits the body through the urethra.

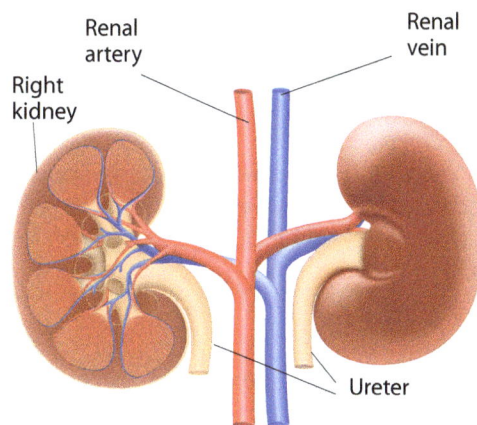

Structure of the kidneys

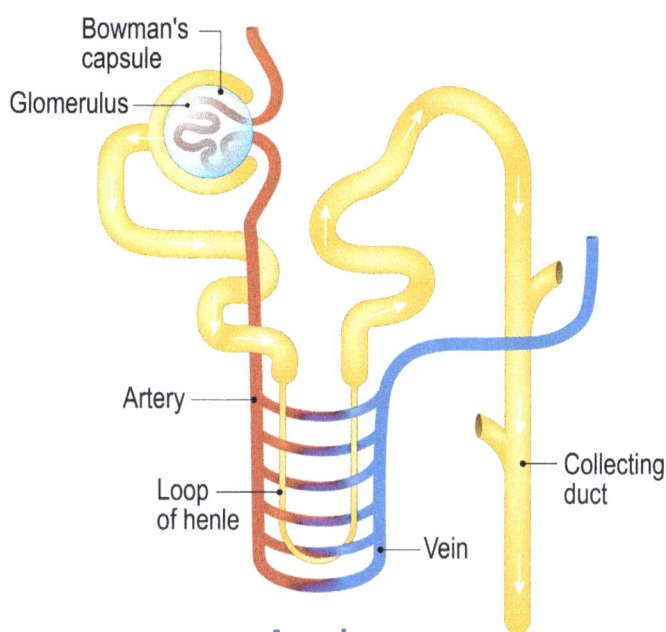

A nephron

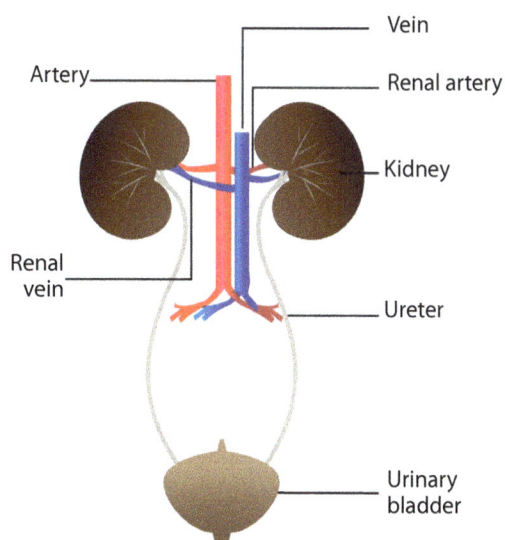

The excretory system

Important terms!

Electrolytes: Ions (such as sodium (Na^+), potassium (K^+) and Calcium (Ca^{2+})) that conduct electricity and play crucial roles in maintaining various physiological functions.

Glomerulus: A sieve-like structure found within the nephron of kidneys.

Ultrafiltration: The process of removing particles from blood in order to produce urine.

Peristalsis: A muscular contraction of the smooth muscle in tube-like organs in order to push substances through them.

Links to other systems

Endocrine system

In addition to the hormones discussed above, the kidneys also produce **erythropoietin** (**EPO**) which stimulates red blood cell production in the bone marrow.

Digestive system

When the body digests proteins, a process called deamination produces **ammonia**. In order to prevent a buildup of this highly toxic substance, the liver converts ammonia into **urea** through a process called the **urea cycle** (also known as the ornithine cycle).

Urea is far less toxic than ammonia and is water soluble, so it can be transported via the blood to the kidneys.

The excretory system complements the digestive system by removing nitrogenous waste (urea) that was not removed through the gastrointestinal tract.

> **Important terms!**
>
> Nitrogenous: Substances containing nitrogen.

• •

Comparative structure and function across taxa

Aves and reptilia

In birds and reptiles, urea does not leave the body via the urethra. Instead, urea leaves their body through the **cloaca**. This is a chamber at the end of the digestive system where the urinary, digestive and reproductive tracts meet. This means that waste and eggs leave the body from the same point.

Birds and reptiles excrete urea in the form of **uric acid** rather than urine. Uric acid is a paste which minimises water loss (allowing reptiles to conserve water in arid environments). This means that birds do not need a bladder, reducing weight and making flying easier.

Mammals

Mammals such as camels and kangaroo rats live in very hot climates and therefore must conserve water to avoid dehydration. These species have evolved longer Loops of Henle within their kidney tubules in the nephrons. A longer Loop of Henle allows for maximum water reabsorption back into the body and the production of concentrated urine. This ensures that these species do not waste any water.

Marine mammals are able to drink salt water as their kidneys have evolved to excrete a much larger amount of salt. This produces very salty, concentrated urine and prevents dehydration.

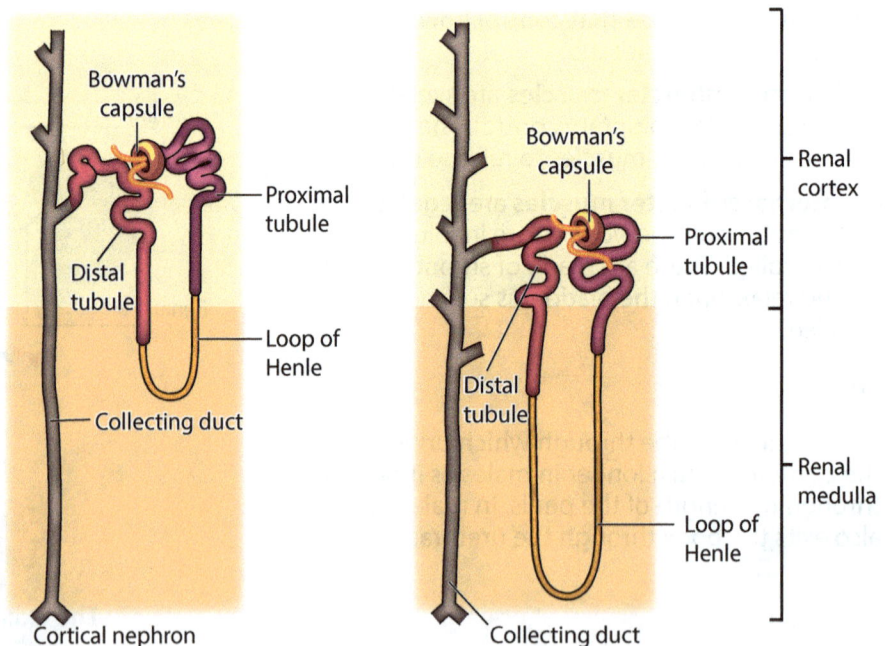

Bowman's capsule — Proximal tubule — Distal tubule — Loop of Henle — Collecting duct — Cortical nephron

Bowman's capsule — Proximal tubule — Distal tubule — Loop of Henle — Collecting duct — Renal cortex — Renal medulla

Adaptation to the loop of Henle in a camel

Aquatic animals

Fish excrete nitrogenous waste as ammonia directly into the water.

Fish also have permeable skin which allows water to pass through it. However, electrolytes (salts), which are essential to life, cannot pass through their skin. This leads to a process called **osmosis**.

Marine (saltwater) fish

- The water surrounding marine fish has a higher concentration of salt than in the fish's body.

- Osmosis causes water to constantly **flow out of the fish's body**.

To maintain osmotic balance, saltwater fish:

- Excrete nitrogenous waste as ammonia directly into the water in **very concentrated urine**.

- Actively excrete excess salts through specialised gills.

Freshwater fish

- The water surrounding freshwater fish has a lower concentration of salt than in the fish's body.

- Osmosis causes water to constantly **flow into the fish's body**.

To compensate for this excess water intake, and maintain electrolyte balance, freshwater fish excrete large volumes of water in **very dilute urine**. They also actively absorb salts from the water through their gills.

Recap Questions

1. What is the role of electrolytes in the animal body?

2. Why does the kidney perform reabsorption when producing urine?

3. Explain ultrafiltration.

4. Name three hormones produced by the kidneys and explain their function.

5. Why does the liver convert ammonia into urea?

6. How does urine production differ between saltwater and freshwater fish?

6.8 The structure and function of the reproductive system in relation to animal physiology

The reproductive system is a vital component of animal physiology, enabling the continuation of species through the process of reproduction. It encompasses a range of structures and functions that differ significantly between males and females, as well as across species, reflecting a wide variety of reproductive strategies and adaptations. This diversity ensures reproductive success in different environments and life histories, from internal fertilisation in mammals to egg-laying in birds and monotremes.

Mammal reproductive systems

Male mammal reproductive anatomy

In all male mammals, the p**enis** is an external organ used for delivering sperm into the female reproductive tract. It is protected by the **prepuce**, a fold of skin which covers and protects the penis when it is not erect. This reduces the risk of damage and infection.

The **urethra** travels through the penis from the bladder and the prostate, allowing both urine and semen to exit the body. The **prostate gland** produces **seminal fluid** which provides a medium for sperm to swim through in the female reproductive tract.

Some species, such as dogs, have a swelling at the base of the penis called the **bulbus glandis**. This expands during mating, preventing the male from withdrawing his penis from the vagina of his mate. This is called the "tie" and can last up to 40 minutes. This increases his chance of successful fertilisation.

Mammals have two **testes** that produce sperm in a process called **spermatogenesis**. They produce the hormone **testosterone** which is responsible for sperm development as well as male secondary sexual characteristics (for example, the mane in male lions). A long coiled tube attached to the back of the testes (the **epididymis**) stores and matures sperm prior to ejaculation.

The sperm travel through a muscular tube called the **vas deferens** from the epididymis to the urethra during ejaculation.

Many species of mammals have a bone in their penis called the baculum. This bone aids the animal in maintaining erection.

Male animals also produce **oestrogen** in their testes and adrenal glands. Although they only produce a very small amount, it is essential to the health and development of bone, the cardiovascular and nervous systems. Oestrogen needs to stay in balance with testosterone to control sex drive, erection and the production of sperm.

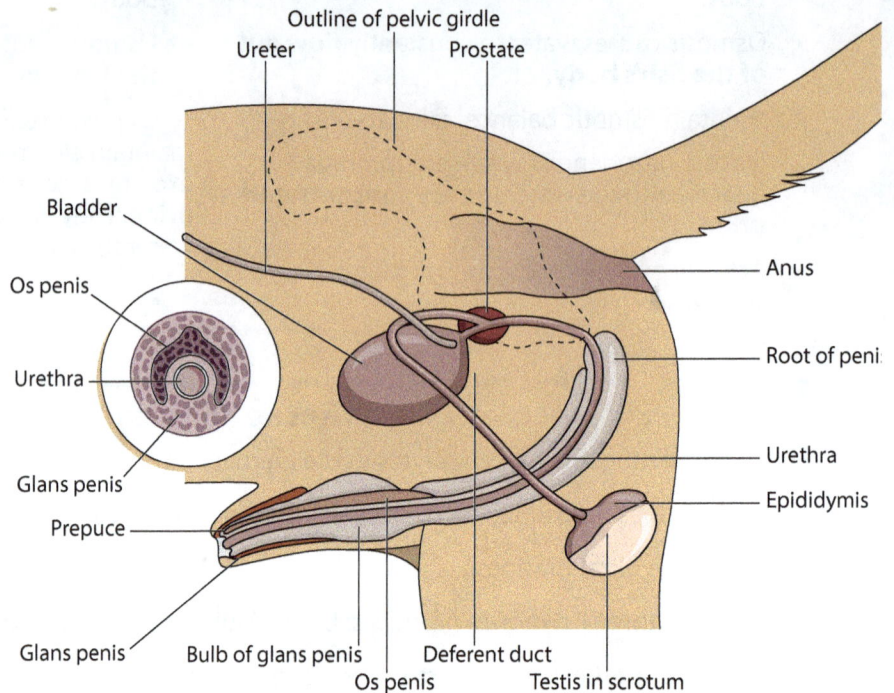

Labels: Outline of pelvic girdle, Ureter, Prostate, Bladder, Os penis, Anus, Urethra, Root of penis, Glans penis, Urethra, Prepuce, Epididymis, Glans penis, Bulb of glans penis, Deferent duct, Os penis, Testis in scrotum

Reproductive anatomy of a male dog

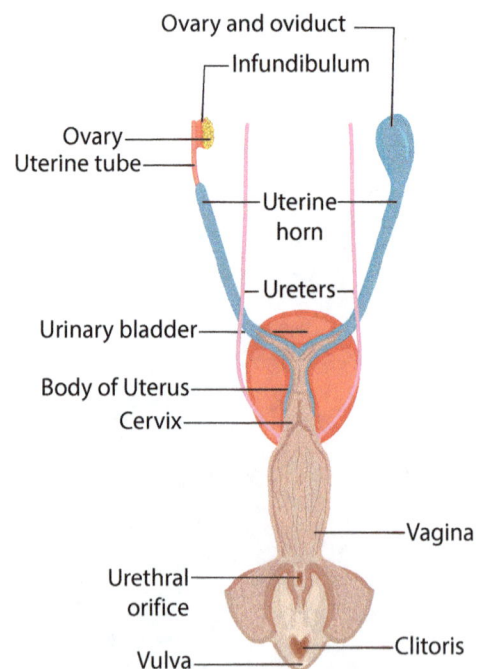

Labels: Ovary and oviduct, Infundibulum, Ovary, Uterine tube, Uterine horn, Ureters, Urinary bladder, Body of Uterus, Cervix, Vagina, Urethral orifice, Clitoris, Vulva

Reproductive anatomy of the female dog

Female mammal reproductive anatomy

Female mammals have two **ovaries** which produce ova (eggs) and the hormones **oestrogen** and **progesterone**. When ovulation occurs, the ovum is released from the ovary and enters the **oviduct**. This is a tube lined with cilia that push the ovum towards the uterus. The oviduct is often the site of fertilisation.

The **uterus** is a small, muscular organ with an amazing capacity to stretch as the foetus grows. The fertilised egg implants into the wall of the uterus where it is able to obtain the necessary nutrients for growth from the mother's body.

At the entrance to the uterus is a sphincter of muscle called the **cervix**. This opens during oestrous to allow sperm entry to the female reproductive system, and again during parturition for delivery of the offspring.

The **vagina** is the canal for both copulation (mating) and parturition (giving birth). It opens to the **vulva**, the only external part of the female reproductive system.

Bird reproductive systems

Bird reproductive systems differ from mammals as they produce hard-shelled eggs rather than giving birth to live young. Females have only one oviduct. This is an adaptation to reduce body weight, making flight easier.

- At the entrance to the oviduct is a funnel-shaped structure called the **infundibulum**. When the egg has been released from the ovary, it is captured by the infundibulum and channelled into the oviduct. In birds, the ovum is attached to the side of the yolk.

- The ovum is fertilised by waiting sperm before travelling through the oviduct to the **magnum** where the egg white (**albumen**) is added around the **yolk**.

- From here, the egg moves into the **isthmus** where fine membranes are added. Next, the egg moves into the uterus where the shell gland excretes a calcium matrix around the egg to produce the shell. This is longest part of the process and can take up to 20 hours.

- When the shell is complete, the egg travels through the vagina to the **cloaca**. This is a chamber that connects the digestive, urinary and reproductive systems in birds. The egg leaves the body via the cloaca.

Very few male birds have a penis. **Ratites** (a group of flightless birds including emus and kiwis) and waterfowl are exceptions. It increases the chances of successful mating. All other species mate using a "cloacal kiss" where male and female birds press their cloaca together for sperm transfer.

> ### Important terms!
> **Copulation**: The act of mating.
> **Parturition**: Giving birth.
> **Sphincter**: A ring of muscle.
> **Ova**: Eggs (ova is plural, ovum is singular)
> **Baculum**: A bone found in the penis of some mammals that aids erection.

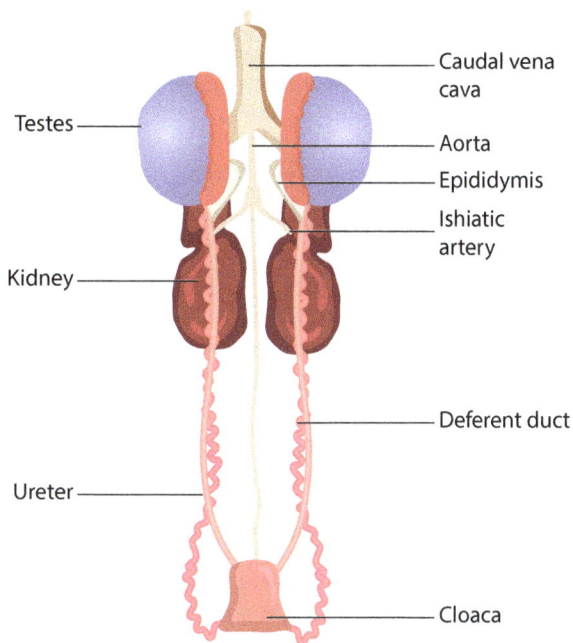

Reproductive organs of a male bird

Labels: Caudal vena cava, Testes, Aorta, Epididymis, Ishiatic artery, Kidney, Deferent duct, Ureter, Cloaca

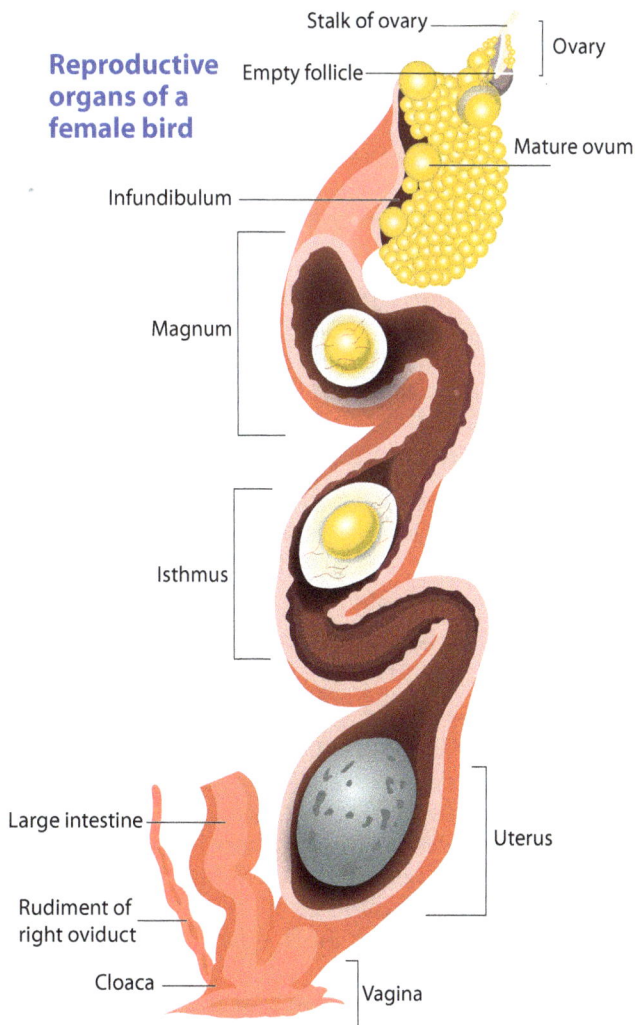

Reproductive organs of a female bird

Labels: Stalk of ovary, Ovary, Empty follicle, Mature ovum, Infundibulum, Magnum, Isthmus, Large intestine, Uterus, Rudiment of right oviduct, Cloaca, Vagina

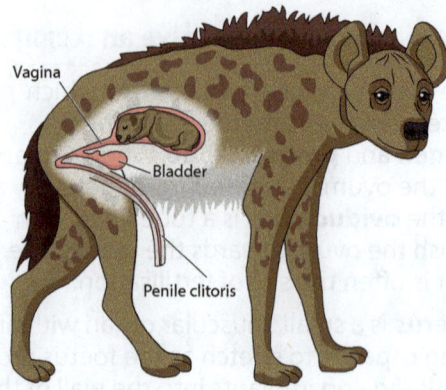

The pseudo-penis of a hyena

Reproductive system adaptations

Male cats have a **barbed penis** which scrapes the female's vagina during mating. This scraping action causes the female to ovulate. Cats are an example of **induced ovulators**. This means every time the female mates, the eggs she releases will be fertilised by that male's sperm. As a female cat may mate with several males, her litter of kittens can therefore have multiple fathers.

Whales and **dolphins** have a retractable penis to allow their body to remain streamlined in the water. These species are known to mate for pleasure as well as reproduction.

Pigs have a corkscrew-shaped penis which matches the sow's coiled vagina.

Snakes have two **hemipenes** (paired penises) which sit inside the tail. During mating, the hemipenis used will be determined by how he is coiled against the female.

Female hyenas have a **pseudo-penis** (elongated clitoris) which make both reproduction and birth challenging. In order to mate, the male must insert his penis into

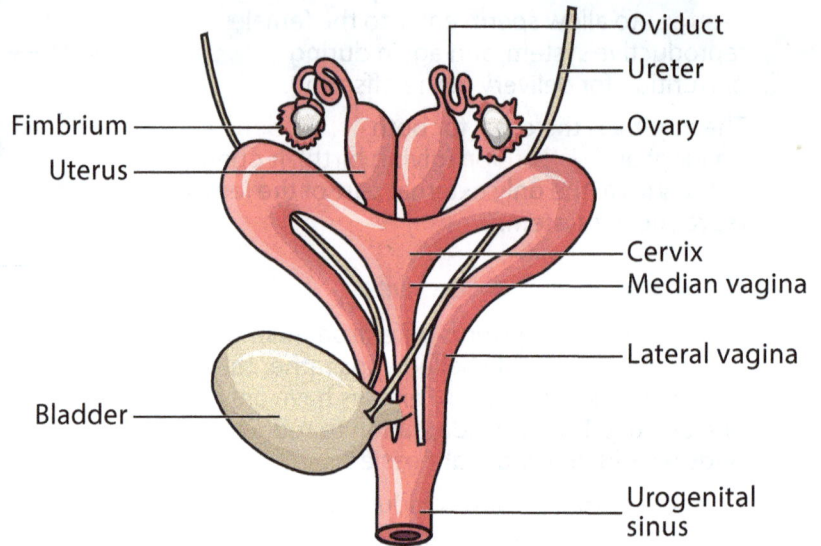

The reproductive organs of marsupials

her pseudo-penis. The urethra connects to the vagina meaning the hyena must give birth through this narrow channel. This is a very painful and often dangerous adaptation, as the urethra has not adapted to stretch in the same way the vagina does, meaning that it can often be damaged.

Marsupials and monotremes

Marsupials have a very different reproductive system to placental mammals. In female, the vagina splits into two lateral (side) vaginas and one median (central) vagina.

- When the female has mated, sperm travels up the lateral vaginas to the two uteri.

- When the joey (infant) is born, it travels down through the median (central) vagina. The joey is born in a very early stage of development and must immediately crawl through the mother's fur into her pouch, where it will attach to one of her nipples until it is fully developed. The mother's body will produce milk that changes with the need of the growing offspring.

The mother can delay the development of the embryo until environmental conditions are favourable. This is known as **embryonic diapause**. This allows the mother to have multiple offspring in the pouch at different stages of development.

To aid the mating process, male marsupials have a **bifurcated penis**. The head of the penis has two prongs to deliver sperm into both lateral vaginas.

Monotremes are a small group of mammals that lay eggs but also have mammary glands for feeding their young. Examples of monotremes are the platypus and the echidna.

Oestrous cycle

The **oestrous cycle** refers to the cycle of physiological changes that occur in female mammals to prepare them for reproduction. There are four stages to the oestrous cycle, each regulated by hormonal interactions between the brain, ovaries and reproductive tract.

> **Remember!**
> An animal 'on heat' is known as being 'in oestrus'. Notice the subtle difference in spelling from 'oestrous cycle'!

Oestrous Cycle

Proestrus
The first stage of the oestrous cycle is triggered by an **increase in follicle stimulating hormone (FSH)** from the pituitary gland. When the ovaries receive FSH, follicles in the ovary begin to develop into ova. **Oestrogen levels rise** as the follicles mature, prompting the female to demonstrate mating behaviours.

Oestrus
During the second stage of the oestrous cycle, the female becomes sexually active and is ready to mate. Ovulation occurs during or just after this phase due to a **surge in luteinising hormone (LH)**. Some species (such as cats and rabbits) are induced ovulators and will only ovulate if they are mated.

Metoestrus
After ovulation, the ruptured follicle forms the **corpus luteum** which **secretes progesterone** to prepare the uterus for pregnancy. This causes the uterine lining to thicken ready to receive the fertilised embryo. If the animal fails to fall pregnant, **prostaglandins are released** from the uterus to break down the corpus luteum, stopping the production of progesterone. This causes the uterine lining to break down and the animal will enter anoestrous. The majority of animals reabsorb this lining rather than bleeding, to reduce the risk of being tracked by predators.

Dioestrus
If the embryo successfully implants into the uterine lining, the animal becomes pregnant. Throughout the pregnancy, **progesterone levels remain high** until the animal is ready to give birth.

Anoestrus
If the pregnancy fails or the animal has given birth, she will now enter anoestrous. This phase of the oestrous cycle sees a **reduction in oestrogen and progesterone levels** to give the reproductive tract time to repair in preparation for the next cycle. During this phase, the female will not be receptive to the male.

Sexual reproduction

Sexual reproduction is vital for animals to pass on their genetics in a way that increases diversity. Animals pass on a random half of their genes to the next generation in a process that allows species to evolve.

Copulation is the physical act of mating, where sperm is deposited in the female reproductive tract. A single sperm cell will **fertilise** the ovum creating a **zygote**, a single cell containing a unique genetic code.

The zygote travels through the uterine tubes into the uterus where it **implants** into the uterine wall to establish a connection with the mother's blood supply. The zygote will grow and develop into a **foetus**, obtaining nutrients and oxygen through its connection with the mother. The length of **gestation** (pregnancy) depends on the species. For example, dogs are pregnant for 63 days whilst elephants are pregnant for 22 months.

Parturition (giving birth) is initiated by a cascade of hormones. When the animal is ready to give birth, **progesterone** levels fall which creates a surge of **oxytocin**. This causes uterine contractions which causes a release of **prostaglandin**. This causes a new wave of contractions which cause a further release of oxytocin. This cycle continues until the contractions are strong enough to push the offspring out of the uterus via the vagina.

Oxytocin and **prolactin** stimulate the production of milk in mammary glands, whilst oxytocin causes the mother to bond with her offspring.

Human influence on breeding

Humans have been selectively breeding animals for thousands of years. Traits such as productivity, temperament and appearance are specifically chosen depending on the need. Modern animal breeding now allows us to use **biotechnological interventions**, including the use of hormones.

Hormone injections are used to control and manipulate reproductive processes in animals. Hormones can be used to influence ovulation, oestrous cycle synchronisation and fertility.

Ovulation can be induced using **gonadotropin-releasing hormone** (GnRH). This is used to ensure that the animal ovulates at the ideal time for artificial insemination.

Super ovulation can be induced in animals using **follicle stimulating hormone** (FSH). This causes the female to release more ova than she would do naturally. These ova can then be collected,

fertilised and implanted into other females who carry them throughout the pregnancy. This is used if a farmer wishes to use the genes from one particular female to produce more offspring than she could do naturally.

Farmers working with large herds of animals often use hormones to synchronise the oestrous cycle of all the breeding females. Hormones such as **progesterone** or **prostaglandins** are used to bring all females into heat at the same time, meaning they will all fall pregnant at the same time. Farmers use this to control when offspring are born, making management of the birthing process much easier. This is commonly used in cattle, sheep and goats.

Important terms!

Copulation: The physical act of mating.

Fertilisation: The joining of the male and female gametes (sperm and ova).

Zygote: A single cell containing the DNA from sperm and egg.

Implantation: The zygote attaches to the uterine lining.

Foetus: The growing embryo becomes a foetus once the major organ systems have formed.

Gestation: The time during which the offspring are developing in the uterus of a mammal.

Parturition: The process of giving birth.

Oestrus cycle synchronisation: Using hormones to bring all animals in a herd into oestrus at the same time in order to facilitate easier management of the breeding process.

Artificial insemination: The act of inserting sperm into the female reproductive tract so that she falls pregnant without actually mating.

Superovulation: Using hormones to cause an animal to produce more ova than she would do naturally.

Recap Questions

1. What is the function of the epididymis in the male reproductive system?

2. Where does fertilisation typically occur in the female reproductive system?

3. Which two hormones are produced by the ovaries and what are their roles?

4. Why have the majority of male birds evolved to lose the penis?

5. How does the barbed structure of the cat's penis aid in reproduction?

6. What is the purpose of the marsupial's pouch in their reproductive strategy?

7. Why do snakes have two hemipenes rather than a single penis?

8. Name each stage of the oestrous cycle in order.

9. During which phase of the oestrous cycle does the corpus luteum form and what hormone does it produce?

10. What is the role of luteinising hormone?

11. What is the difference between fertilisation and implantation?

12. Which hormones are responsible for uterine contractions during parturition?

6.9 The structure and function of the integumentary system in relation to animal physiology

The **integumentary system** is a complex and vital organ system that serves as the outermost protective layer of an animal's body. It includes the skin, hair, scales, feathers, claws, hooves and other specialised structures, which vary across different animal groups. This system acts as the **first line of defence** against environmental threats, such as pathogens, physical injuries, and extreme temperatures, while also playing a critical role in sensation, thermoregulation and communication.

The structure and function of the integumentary system

Key functions of the integumentary system include:

- **Protection:** Forms a physical and chemical barrier to shield the body from harm.

- **Sensation:** Houses sensory receptors that detect touch, pain, heat, cold and other stimuli.

- **Thermoregulation:** Maintains body temperature through structures like sweat glands and fur.

- **Water balance:** Helps prevent water loss or gain in varying environments.

- **Wound healing:** Participates in tissue repair and recovery.

- **Specialisation:** Adapts to specific needs in different species, such as feathers for flight or scales for water resistance.

Epidermis

The **epidermis** is the outer layer of skin. Its primary function is to protect the animal body from physical damage and pathogens. The epidermis is formed from stratified squamous epithelial tissues. The outermost layer is made of dead, flattened keratinised cells which create a tough, waterproof layer.

Dermis

The **dermis** is the second layer of skin and it sits beneath the epidermis. This layer of skin contains **blood vessels**, sensory nerve receptors and glands.

The sensory nerve cells detect stimuli such as touch, pain, heat and cold, enabling the animal to interact with its environment.

Within the dermis, there are two types of glands;

- **Sebaceous glands** produce an oily substance called **sebum** which keeps the skin and hair soft and pliable, whilst helping to waterproof and protect the skin from microorganisms. These are found in all areas of the body except the palms and soles of the feet.

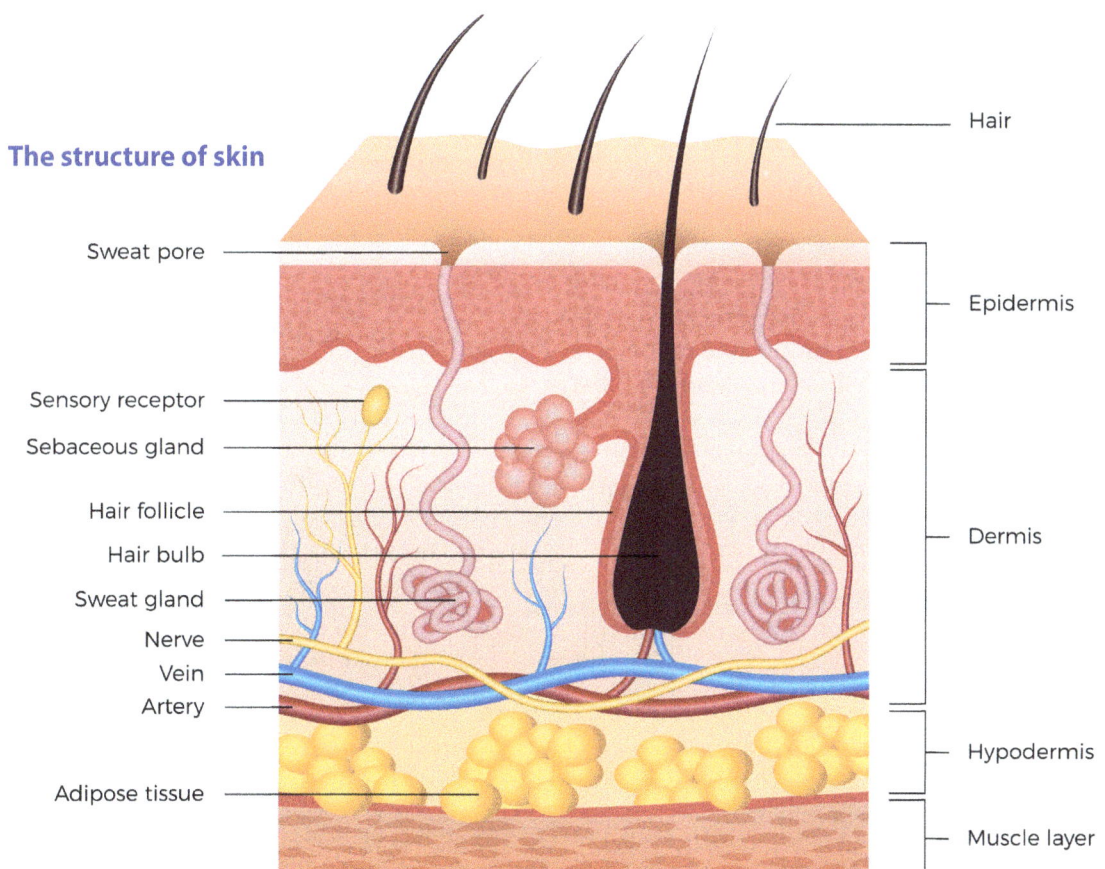

The structure of skin

Labels (top to bottom, left side): Sweat pore, Sensory receptor, Sebaceous gland, Hair follicle, Hair bulb, Sweat gland, Nerve, Vein, Artery, Adipose tissue

Labels (right side): Hair, Epidermis, Dermis, Hypodermis, Muscle layer

- **Sweat glands** play a major role in thermoregulation and, in some species, communication. **Eccrine sweat glands** produce watery sweat that cools the body through evaporation. This also helps to remove small amounts of metabolic waste such as urea and salts. **Apocrine sweat glands** produce a thick sweat that is broken down by bacteria producing odour. In some species of animal, these glands are involved in pheromone communication.

Hypodermis

The **hypodermis** is the deepest layer of skin and is responsible for wound healing and anchoring hair follicles. The hypodermis acts as a physical barrier offering protection and insulation.

Role of the integumentary system as the primary defence mechanism in immunology

The **integumentary system** plays a crucial role as the **first line of defence** in the immune system. Its structure and components provide both physical and chemical barriers against pathogens, as well as initiating immune responses when the skin is breached. These defensive mechanisms are essential for maintaining homeostasis and protecting the body from infections and injuries.

Physical barrier

The integumentary system provides a tough, flexible shield that prevents pathogens entering the body and reducing the risk of damage to internal organs.

- The epidermis provides a tough, waterproof outer layer which prevents pathogens entering the body.
- As dead skin cells shed, they take potential contaminants with them.
- The skin also provides protection from physical damage that animals may experience from their interactions with their environment and other animals.

Chemical barrier

As added protection against pathogens, the integumentary system produces a range of chemical substances that neutralise or inhibit pathogen growth.

- Sebum produces an oily, acidic layer on the skin which reduces bacterial and fungal growth.
- Sweat contains antimicrobial peptides which actively kill bacteria and fungi.

Cellular defence

There are **specialised immune cells** in the skin that detect and respond to pathogens that have entered the body through wounds.

- **Langerhans cells** found in the epidermis capture and process antigens, presenting them to T-cells to activate the immune system.
- **Mast cells** in the dermis release histamines and other substances promoting inflammation and attracting white blood cells.
- **Macrophages** in the dermis engulf pathogens or debris to help clear infections.

Wound healing

When the skin is damaged, the integumentary system triggers immediate repair mechanisms.

- **Platelets** in the blood form a clot to seal the wound and prevent pathogen entry.
- **Inflammation** attracts immune system cells to the site of damage, to fight infection and remove damaged tissue.
- Fibroblasts in the dermis produce new collagen and elastin, substances that are used to rebuild the skin, whilst there is a proliferation of keratinocytes to restore the epidermis.

> **Important terms!**
>
> Stratified squamous epithelial tissues: Layers of squamous cells that form the surface layers on epithelial tissues.
>
> Keratinised: Cells that contain keratin, a protein found in skin and hair.

Bleeding

Blood clot

Inflammatory

Scab

Fibroblast

Macrophage

Blood vessel

Proliferative

Fibroblasts proliferating

Subcutaneous fat

Remodeling

Freshly healed epidermis

Freshly healed dermis

Wound healing

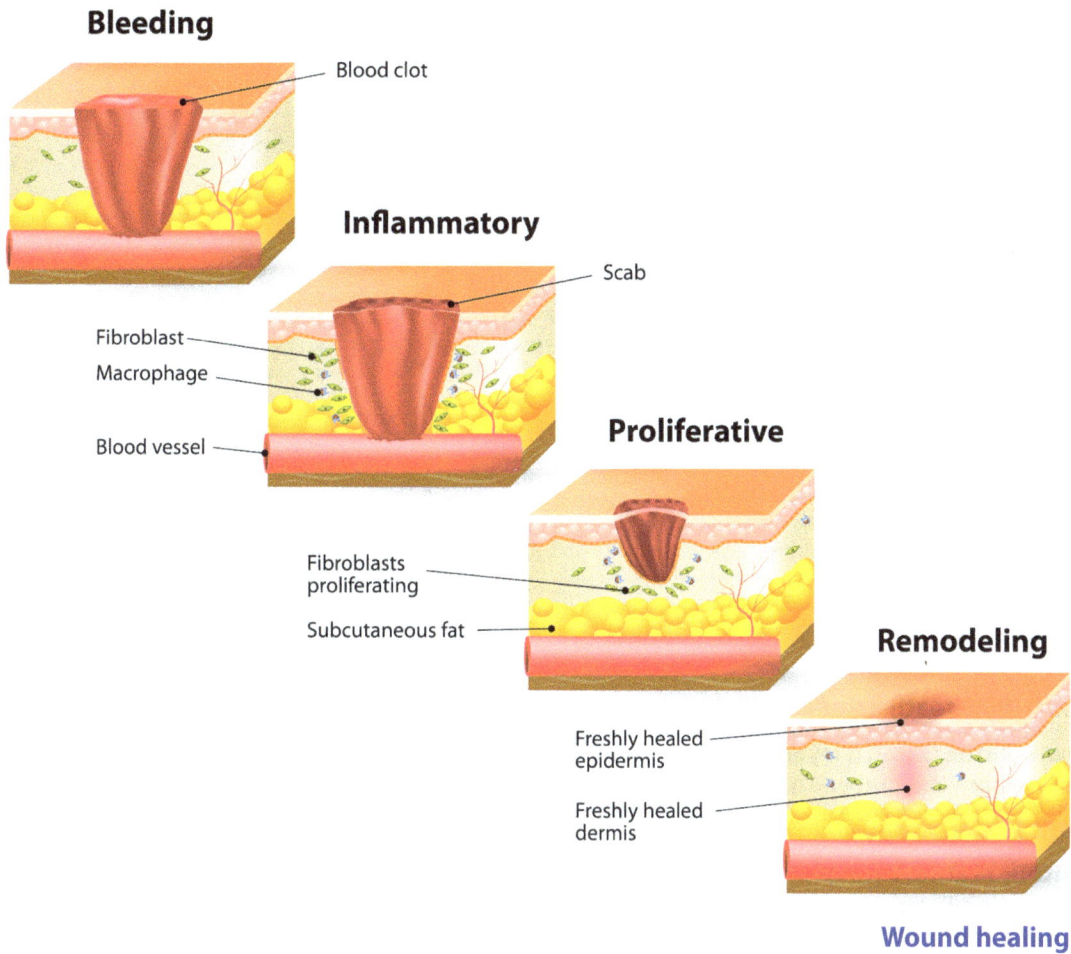

Comparison of the structure and function of the following components of different taxa

Animals have evolved diverse adaptations to the integumentary system to suit their environment and lifestyles. Specialist structures provide animals with the means to survive in very specific environments.

Aves

Birds have evolved a variety of bill (beak) shapes and structures. In all cases, the bird's bill enables them to exploit a specific food source in their environment. The bill is formed from two bony projections, the upper and lower mandibles, which are covered in a keratinised layer called the **rhamphotheca**.

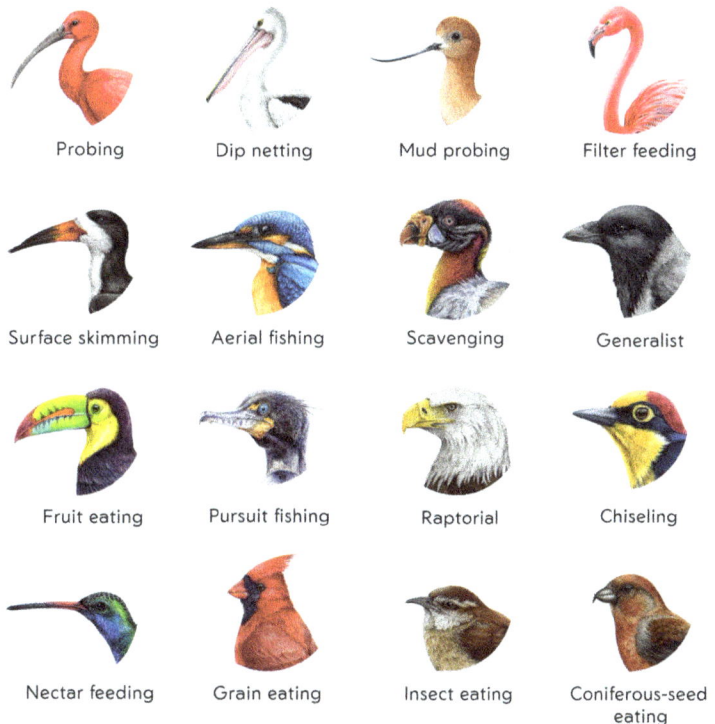

Probing

Dip netting

Mud probing

Filter feeding

Surface skimming

Aerial fishing

Scavenging

Generalist

Fruit eating

Pursuit fishing

Raptorial

Chiseling

Nectar feeding

Grain eating

Insect eating

Coniferous-seed eating

Bird beaks and their functions

129

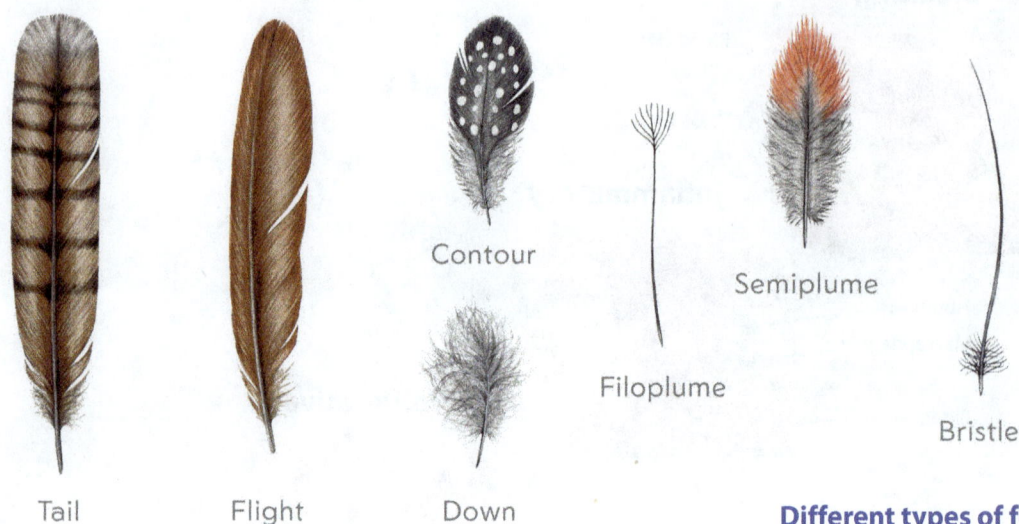

Contour

Semiplume

Filoplume

Bristle

Tail

Flight

Down

Different types of feathers

Birds have evolved different types of feathers, each for a different purpose.

- **Wing/flight feathers** enable flight.
- **Down feathers** trap a layer of air against the skin keeping the bird warm.
- **Tail feathers** give the bird control in the air.
- **Contour feathers** are arranged in overlapping layers. The tips of these feathers are waterproof, providing the bird with a protective outer layer.
- **Semiplumes** provide an additional layer of insulation.
- **Filoplumes** are very simple feathers with few barbs. They provide sensory information about temperature, wind speed and feather movements to help birds keep their feathers in order.
- **Bristle feathers** are often found around the bird's eyes and beak provide protection from dust and insects, much like eyelashes in other species.

Birds have a specialised gland near the base of the tail called the **uropygial gland**. This produces oils.

Fish

One of the biggest adaptations to the fish integumentary system are the scales that cover the body. **Scales** are a transparent, colourless layer that provide protection and streamlining for swimming. Many fish species have a mucus layer which covers the scales, producing a barrier against pathogens and parasites.

Although the **gills** are primarily a respiratory structure, they are part of the integumentary system. The thin membranes enable gas exchange.

Fins are supported by rays or spines and assist in locomotion and stability. In some species, fins

have venomous spines providing protection from predators.

Although the scales of a fish are transparent and colourless, the skin under them is often very colourful. Many species of fish are very well camouflaged making them excellent ambush predators.

Chromatophores are colour-changing cells that allow certain species to blend with their environment, mimic other species and put on displays to attract mates.

A lionfish with venomous spines

A well-camouflaged fish

Reptilia

Reptiles have evolved a dry, keratinised integumentary system, specially adapted for life in hot, arid climates. Many species have scales composed of keratin as the outer layer of their epidermis. Scales provide physical protection as well as preventing water loss.

Species of reptiles such as crocodiles have evolved bony plates called **osteoderms**. These are another type of scale but these are formed in the dermis. Again, their function is to provide protection from physical damage and water loss.

Unlike mammals and birds, reptile skin does not flake away as cells die. **Ecdysis** (shedding) takes place periodically and involves the animal completely shedding the outer layer of their epidermis. This process removes parasites, enables animals to grow and helps to remove damaged skin. Snakes and geckos shed their skin in one continuous piece whilst lizards tend to shed in smaller patches.

Reptiles have evolved different colours and patterns based on where they live in the environment. Being able to blend into the surroundings means predators are able to ambush prey whilst prey species are able to hide.

Some species have evolved the ability to change colour for communication and thermoregulation. In many species, this ability is fairly limited (such as the darkening of the beard in bearded dragons). However some species, such as the chameleon, have an impressive colour palette.

Amphibians

Amphibians have a moist, permeable integumentary system which has evolved to allow gas exchange. They lack scales and their skin is soft and thin, allowing gases to diffuse through the skin directly into the blood supply. The skin must remain moist at all times to enable gas exchange.

Amphibians produce mucus which covers their skin. This mucus helps to keep the skin moist, aids in gas exchange and prevents the skin cracking if it does become too dry. In many species, the mucus covering the skin is toxic, acting as an excellent defence against predators.

An Amazon tree boa snake shedding its skin

A well-camouflaged lizard

A colour-changing chameleon

European common brown frog

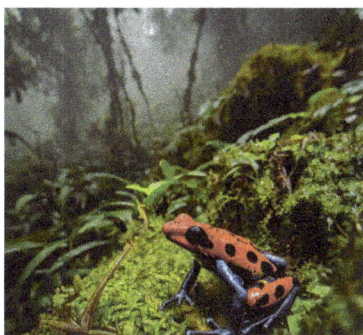
A poison dart frog

6 Anatomy and physiology

Exoskeletons

Invertebrates

Unlike vertebrate species, many invertebrates' integumentary system has evolved into a hard **exoskeleton** which provides support and protection. Exoskeletons are primarily made of **chitin**, a tough, flexible polysaccharide which is reinforced with calcium carbonate in some species (such as crabs and lobsters). The exoskeleton provides a rigid framework for muscle attachment, enabling efficient movement.

Arthropods and other invertebrates must shed their exoskeletons as they grow. This process is called **ecdysis** and is controlled by hormones. It leaves the animal very vulnerable to attack.

Some invertebrate species, such as molluscs, have a hard shell made from calcium carbonate which provides physical protection from predators and environmental hazards. These species are often very fragile with soft bodies that are able to retract into the shell when necessary.

Important terms!

Antimicrobial peptides: Small molecules that are produced by the immune system as a first line of defence against bacteria.

Antigens: A toxin or pathogen which triggers an immune response in the body. Antigens very often lead to the production of antibodies.

Histamine: A compound released by cells following injury or in response to an allergen. Histamines cause inflammation, bringing more blood and therefore white blood cells to the affected area.

Fibroblasts: Cells that produce collagen and other fibres in connective tissues.

Keratinocytes: Cells that produce keratin.

Proliferation: A rapid increase in the number of something.

Chitin: A tough, flexible polysaccharide that forms exoskeletons in many invertebrates.

Polysaccharide: A complex molecule made from chains of sugars.

Calcium carbonate: A white, insoluble solid with the formula $CaCO_3$. Found commonly in chalk and limestone as well as forming mollusc shells and corals.

Arthropods: An invertebrate animal of the phylum *Arthropoda* such as insects, spiders and crustaceans.

Recap Questions

1. Name the three layers of the skin and their functions.

2. What is the difference between apocrine and eccrine sweat glands?

3. What is the function of sebum?

4. How does the skin heal itself when damaged?

5. How does skin act as a physical barrier?

6. How does the skin act as a chemical barrier?

7. Name three types of feather and their function.

8. What is the purpose of the uropygial gland in birds?

9. How do reptile and amphibian skin differ?

Practice Questions

1. Animals have evolved different digestive systems. Discuss which of these is the most efficient and why. (6 marks)

2. Explain why animals have different length digestive tracts linked to their diet. (6 marks)

3. Discuss three differences in the respiratory systems of mammals, birds and amphibians. (6 marks)

4. Explain the differences between aerobic and anaerobic respiration. (4 marks)

5. Describe how the heart structure is related to its function. Your answer should include how the heartbeat is generated. (4 marks)

6. Describe the role of the circulatory system in immunity. (3 marks)

7. Discuss the function of two different hormones produced by the endocrine system. Assess their impact on homeostasis. (6 marks)

8. Explain how bones, muscles, tendons and ligaments work together to produce movement. (6 marks)

9. Discuss different adaptations of the pentadactyl limb. How do these adaptations enable animals to survive in their environment? (8 marks)

10. Label the bones indicated on the dog skeleton below. (3 marks)

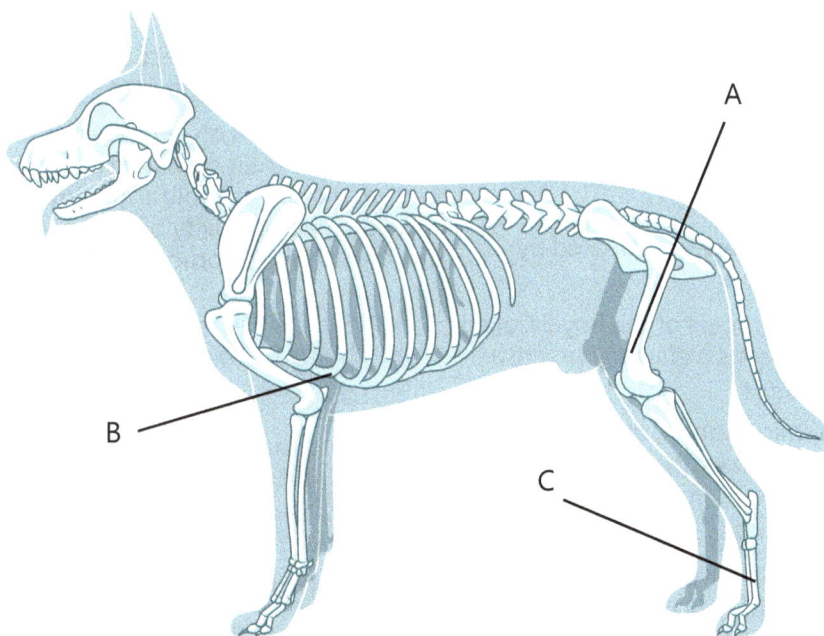

11. Assess how animals use different senses in order to survive. Your answer should include one prey species and one predator. (6 marks)

12. Explain the role of the sympathetic and parasympathetic nervous systems. (4 marks)

13. Desert dwelling species are able to conserve water to enable survival in arid climates. Explain the kidney adaptations that make this possible. (4 marks)

14. Explain how the body produces urea and describe its excretion, highlighting the roles of the liver, kidneys and bladder. (5 marks)

15. All birds lay hard-shelled eggs in order to reproduce. Explain each stage of this process. (4 marks)

16. Explain what happens in each stage of the oestrus cycle. (8 marks)

7.1 Categories and techniques of animal medication

There are two main categories of animal medication.

- **Prescribed medication** – issued on prescription by a veterinary surgeon (vet)

- **Non-prescribed medication** – no prescription is needed for these medicines

There are many different medicines used in animal care that are specific to disorders, diseases and parasites or specific to the animal being treated.

Routes of administration of medication (techniques) will depend on the animal as well as the type of medication.

Legislation

There are legal requirements to be considered when using animal medication for any animal.

Animal Welfare Act 2006

- Does not have detailed rules about the specific use of medication.

- Duty of care does require animals to receive necessary veterinary treatment, including medication to prevent unnecessary suffering.

- Works alongside others; Veterinary Surgeons Act 1996 & Veterinary Medicines Regulations 2013.

Veterinary Surgeons Act 1966

- Governs the practice of veterinary medicine and surgery.

- Ensures that animals are only treated by those qualified to do so.

- Only registered members of the Royal College of Veterinary Surgeons (vets) may practise veterinary surgery.

- Restricts the administration, dispensing and prescribing of veterinary medicines to those who are qualified to do so.

- There are some exceptions for the prescription of veterinary medicines by other professionals. For example in certain cases veterinary nurses may do so under the direction of a vet.

- It works alongside other laws and regulations to control the sale and use of veterinary medicines, requiring appropriate record-keeping and responsible use of medications.

Veterinary Medicine Regulations 2013 (VMR)

- Main regulations controlling animal medication.

- Covers any substance or combination of substances that treats or prevents ill health in animals.

- Controls the use, distribution and administration of veterinary medicines and medicated feed.

- Designed to protect animal health, public health and the environment.

- The controls on veterinary medicines include their manufacture, advertising, marketing, supply and administration. It is the responsibility of anyone engaged in these activities to comply with the VMR.

Animal Welfare (Licensing of Activities Involving Animals) (England) Regulations 2018

- Established a framework for licensing businesses and individuals involved in various activities with animals, ensuring that animal welfare is prioritised.

It states that:

- Prescribed medicines must be stored safely and securely to safeguard against unauthorised access, at the correct temperature, and used in accordance with the instructions of the veterinarian.

- Medicines other than prescribed medicines must be stored, used and disposed of in accordance with the instructions of the manufacturer or veterinarian.

Medication

The main authorised veterinary medicines are:

- **POM-V** Prescription-only Medicine – can only be prescribed by a vet following a diagnosis.

- **POM-VPS** Prescription-only Medicine – can only be prescribed by a vet, pharmacist, suitably qualified person (SQP). The animal need not be under their care.

- **NFA-VPS** Non-Food Animal – does not have to be prescribed. However, can only be supplied by a vet, pharmacist, suitably qualified person (SQP) for non-food animals, e.g. horses and pets.

- **AVM-GSL** Authorised Veterinary Medicine – can be sold directly to the public with no need for a prescription by any retailer.

Examples of medicines and uses

POM-V	All antibiotics and hormone products, anti-inflammatories, antivirals, some vaccines, certain classes of wormers (endoparasiticides), anaesthetics.
POM-VPS	Many wormers and ectoparasiticides (flea & tick) treatments, local anaesthetics.
NFA-VPS	Mainly endoparasiticides and ectoparasiticides
AVM-GSL	Medicines deemed extremely safe, includes shampoos and other products with very small amounts of active ingredient.

Situations and scenarios requiring medication

- Most domesticated animals will receive medication from their vets after a clinical consultation due to illness, accident or injury. Preventative medicines such as vaccinations and parasitic treatments are issued by a vet on a regular basis. For example, annual vaccinations and monthly worming treatments.

- Some medication can be purchased in pet stores or by those supplying animal products. This will be AVM-GLS and some NFA-VPS medication, for example flea collars and some wormers. It may be that a SQP is required to sell these items, such as flea and tick treatments.

- Animal collections that contain wild captive animals will have either an onsite vet or be associated with a veterinary surgery which specialises in wild captive animals. They will treat these animals with medication due to illness, accident or injury.

- Wild animals tend to only be given medication if they are presented at a veterinary surgery or wildlife rescue centre. It can be stressful for the wild animals to be treated by humans and decisions must be taken as to what is best for the welfare of the animal. Wild animals may become injured in road traffic accidents, become caught in fences, attacked by predators or domestic animals such as dogs or cats. Wild animals that could require treatment include; birds, rodents, hedgehogs, rabbits, hares, frogs & toads, foxes, badgers, otters, deer, seals and other sea mammals.

Hedgehogs are often injured in traffic accidents

Processes in managing and storing veterinary medicines

It is essential to manage and store medicines correctly, for legal purposes and to ensure that the medication can be effective.

Types

There are many medicines used in veterinary medicine. The table on the next page summarises the main types and their use.

Types of medication	Used for
Antibiotics	Used to treat or prevent bacterial infections.
	Can be in a variety of forms: liquid, tablet, spray etc.
Ectoparasiticides	Used to treat and prevent ectoparasites infections (parasites that live on an animal) e.g. fleas, tick, mites and lice
	Can be in a variety of forms: tablets, spot on treatments, sprays, collars and shampoos
Endoparasiticides	Used to treat and prevent endoparasitic infections (parasites that live in an animal) e.g. roundworms and tapeworms. Can also include hookworms, whipworms, lungworms and liver flukes.
	Can be in a variety of forms: tablets, injections, spot on treatments, liquid etc.
Vaccinations	Preventative medication that can require boosters at regular intervals to prevent infectious diseases from affecting an animal.
	It is usually given as a subcutaneous injection. For example, in cats and dogs into the loose skin at the neck. In dogs, the kennel cough vaccination is given via a nasal spray.
Anaesthetics	General anaesthetics cause the animal to become unconscious for a period of time. The animal experiences muscular relaxation (no movement) and no pain sensations. Used for operations or when the animal needs to be intubated or lie still, such as during x-rays, CT or MRI scans.
	Local anaesthetics numb a localised area of skin or a tooth, usually via an injection. Enables a localised area to be treated or repaired with stitches.
Anti-inflammatories	Anti-inflammatories for animals include corticosteroids and nonsteroidal anti-inflammatory drugs (NSAIDs):
	• Corticosteroids
	These are commonly used in cats and dogs to prevent or suppress inflammation. However, they can also suppress the immune response, which may increase the risk of infections.
	• NSAIDs
	These are used to treat pain, inflammation and high temperatures in animals. They can be administered by injection, tablet, capsule, as liquids, or as eye drops.
Antifungals	Used to treat fungal infections.
	Mainly topical treatments. May require clipping of fur or nails for adequate exposure.
	Can also be in the form of creams as well as shampoos.
Anti-emetics/emetics	Anti-emetics are used to prevent or decrease vomiting caused by an underlying condition or toxic substance
	Emetics are used to induce vomiting in animals that have ingested some sort of toxin (including objects). They are used mostly in emergency situations and are most effective when used soon after ingestion.
Analgesics	These are medications that can provide pain relief for animals.
	Can be used to treat acute pain, chronic pain or pain after surgery.
	Can also be given before treatment.
	Includes NSAIDS, opioids and local anaesthetics.
Antiviral	The aim of antiviral therapy is to remove the virus, while minimally impacting the animal, and to prevent further viral invasion.

Controlled drugs storage

- **Controlled drugs** (**CDs**) are very commonly used in veterinary practice.
- They need to be strictly managed.
- Regulation of CDs is enforced by the Home Office, the Veterinary Medicines Directorate (VMD) and the Royal College of Veterinary Surgeons (RCVS).
- Misuse of Drugs Act (MDA) 1971 controls the availability of drugs that are considered dangerous or otherwise harmful. CDs used in veterinary practice are allowed to be possessed, prescribed, administered or supplied by vets under this act.
- Misuse of Drugs Regulations 2001 classifies CDs into 5 schedules and states whether the drugs must be kept in 'safe custody' or not. Schedule 1 CDs are not used in veterinary practice. Schedule 2 CDs require safe custody (with one exception - quinalbarbitone). One Schedule 3 CD – buprenorphine has to be kept in safe custody.
- Safety custody is a Controlled Drugs Cabinet. Cabinets must adhere to the Safe Custody Regulations 1973 in terms of design and construction.

Other requirements include the following:

- The cabinet must be bolted to the wall or floor.
- Preferably double locked (with separate keys).
- Cabinets must be kept locked when not in use.

Figure 7.1.2 A lockable medicine cabinet

- The lock must be different to any other lock on the premises.
- Keys must only be available to vet and authorised members of staff.
- The cabinet should be for the sole use of storing CDs.
- The cabinet must not have anything attached to it which identifies it as a CD cabinet.
- Ideally, CDs should never be left unattended in a vehicle. However, if this is necessary, there should be a locked container fixed within the boot of the car, which must meet the requirements of the Safe Custody Regulations.
- Examples of veterinary CDs include opioids, ketamine and benzodiazepines.

Suitably qualified person (SQP)

- A person who can legally prescribe or supply certain animal medicines (POM-VPS and NFA-VPS) as per the Veterinary Medicines Regulations 2016.
- They must qualify to be a SQP with training and examinations and be included on an approved register.
- The register is maintained by the Animal Medicines Training Regulatory Authority (AMTRA).
- SQPs work in organisations such as veterinary practices, pet shops, agricultural merchants and equine suppliers.
- There are different SQP categories depending the species they have trained on, from single categories (companion only, farm only, equine only, avian only) to combinations of categories up to those who have trained in all 4 categories.
- Can also be known as Registered Animal Medicine Advisors (RAMAs).

Stock management

As with all stock management, care must be taken with:

- Suitable stock record system
- Temperatures items required to be stored at
- Expiry dates – it is an offence to supply or administer an out-of-date medicine
- Stock rotation - so that shortest expiry dates are used first
- Reducing waste
- Saving money
- Setting minimum and maximum stock levels of each item
- Noticing & removing damaged stock
- Regular stock checks
- Maintenance of a CD register

Records of the retail supply (which includes administration) of POM-V and POM-VPS medicines must be kept for 5 years. These records need to include:

- Date of receipt and supply/administration
- Name and quantity of the VMP
- Name and address of the supplier or recipient
- Batch number

Annual audits of POM-V and POM-VPS medicines should be carried out by vet practices.

The supply and use of controlled drugs (CDs) should be recorded, and a running total kept in a Controlled Drug Register. A system of reconciling the balance in the register with the stock in the CD cupboard should be done regularly (at least weekly).

Techniques used for administering medicines and their suitability for different purposes

Medications can be administered to animals via a variety of routes that may be specific to the medication, animal or situation.

Enteral

Medication is given internally orally via the mouth (PO) or via the rectum (PR) or via a tube to the gastrointestinal tract.

- Can be given in feed, water or tablet form.
- Aim is to deliver the medication directly to the gastrointestinal tract.
- Examples: wormer treatments, antibiotics, dehydration fluids

Parenteral

This is any **internal** route **other** than enteral for medication administration.

There are four main parenteral routes:

2. intravenous (IV) into a vein – for example, anaesthetic drugs

3. intramuscular (IM) into a muscle – for example, antibiotics

4. subcutaneous (SC) under the skin – for example, vaccinations

5. intra-articular (IA) into a joint – for example, drugs for osteoarthritis

In all cases the drug is usually administered via a hollow needle.

Topical

These treatments are applied onto the surface of the body (skin, eyes). This can include creams, lotions, shampoos, pastes and spot on treatments. These can include antiseptic, anti-fungal, anti-inflammatories and parasiticides.

Inhalation

These treatments pass directly into the respiratory system either via the mouth or nose.

For example, asthma treatments in dogs and cats and the intranasal inhalation of one of the kennel cough vaccinations in dogs

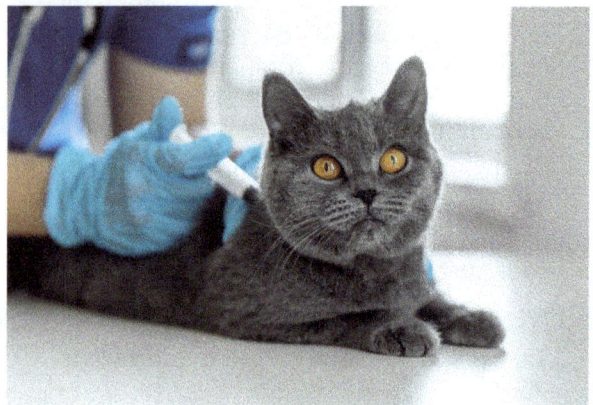

A subcutaneous injection

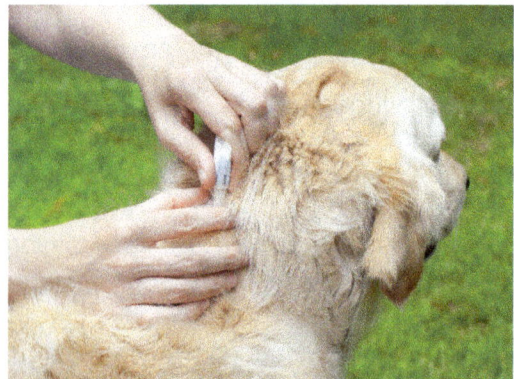

Applying a topical treatment

Inhalation

Medical treatment by non-qualified staff

Legal requirements

The Veterinary Surgeons Act 1966 ensures that animals are treated only by those people qualified to do so.

There are sections that list exceptions to this for non-qualified staff (Schedule 3):

- The animal owner, a member of their household or their employee - may carry out minor medical treatment.

- The animal owner or person engaged in caring for animals used in agriculture, who may carry out medical treatment or minor surgery not involving entry into a body cavity.

- Registered veterinary nurses who may carry out medical treatment and minor surgery (not including entry into a body cavity).

- Student veterinary nurses who may carry out medical treatment and minor surgery (not including entry into a body cavity).

- Veterinary students who are undertaking the clinical part of their course – due to the Veterinary Surgeons (Practice by Students) (Amendment) Regulations 1993.

- Registered farriers – due to the Farriers (Registration) Acts 1975 and 1977.

- Persons providing physiotherapy – due to the Veterinary Surgeons (Exemptions) Order 2015.

- Anyone administering emergency first aid to save life or relieve pain or suffering.

Consequences for unlicensed procedures

The Schedule 3 (section 3) exceptions give details of who and when people other than vets can give medical treatments (as in the previous section).

If an unlicensed procedure breaches Schedule 3 of the Veterinary Surgeons Act 1966, the person who carries it out is guilty of an offence. They can be liable for:

- A penalty fine not exceeding £1000 for offenses not brought before a jury (lesser charges)

- A fine on conviction of an offense at a Crown Court (prosecution of greater charges)

In vet practices, vets need to consider 6 key things (mnemonic SUPERB) when delegating tasks to registered vet nurses (RVN) or student vet nurses (SVN):

SUPERB

Specific procedure - Is the procedure medical treatment or minor surgery, not involving entry into a body cavity?

Under care – is the animal under your care?

Person – can you delegate to this person?

Experience - Does the person feel capable and have sufficient competence and experience?

Risks - Have you considered the risks specific to this case?

Be there - Are you available to direct or supervise, as necessary?

Important terms!

POM-V: Prescription-only medicine – can only be prescribed by a vet following a diagnosis.

POM-VPS: Prescription-only medicine – can only be prescribed by a vet, pharmacist, suitably qualified person (SQP).

NFA-VPS: Non-food animal medicine – does not have to be prescribed, however, can only be supplied by a vet, pharmacist, suitably qualified person (SQP) for non-food animals, e.g. horses and pets.

AVM-GSL: Authorised veterinary medicine – can be sold directly to the public with no need for a prescription by any retailer.

Controlled Drugs: Medication that may be addictive or abused.

SQP: Suitably Qualified Person, also known as Registered Animal Medicine Advisors (RAMAs)

Enteral: Internal route for medication into the body orally or rectally.

Parenteral: Any internal route for medication into the body that is not enteral, e.g. into a vein, joint, under the skin or into muscle.

Topical: External route for medication when applied to the surface of the skin.

Schedule 3: Section of the Veterinary Surgeons Act 1966 that gives the exceptions to the rule that animals are only treated by people qualified to do so – i.e. non-qualified staff.

1. Which laws must be considered when using animal medication?

2. What are the four main types of veterinary medicines?

3. What is the difference between ectoparasiticides and endoparasiticides?

4. What are controlled drugs and how should they be stored?

5. What is an SQP?

6. What is a CD register?

7. What are the four main routes of medicine administration to an animal?

8. Can non-qualified staff treat animals and what can they do?

9. What are the consequences of unlicensed procedures?

7.2 Diseases, disorders, parasites and notifiable diseases that can affect animals

Signs and symptoms of diseases, disorders and parasites

Understanding diseases, disorders and parasites that affect animals is crucial for ensuring their health and welfare. Recognising early signs of health problems in animals is vital for effective management and treatment. Below are some common symptoms.

Common disease symptoms

Respiratory Issues: Coughing, wheezing, nasal discharge, laboured breathing.

Gastrointestinal Problems: Diarrhoea, vomiting, bloating, weight loss.

Neurological Symptoms: Seizures, tremors, disorientation, paralysis.

Dermatological Issues: Hair loss, excessive scratching, sores, scaly skin.

Behavioural Changes: Lethargy, aggression, excessive vocalisation, withdrawal.

Fever and inflammation: Increased body temperature, swollen joints or glands.

Disorders affecting biological systems

Cardiovascular Disorders: Heart murmurs, irregular heartbeat, cyanosis (blue gums).

Digestive Disorders: Constipation, colic, excessive gas.

Musculoskeletal Disorders: Lameness, stiff movements, joint swelling.

Endocrine Disorders: Diabetes, thyroid imbalances (weight changes, coat quality deterioration).

Reproductive Disorders: Infertility, dystocia (difficulty giving birth), hormonal imbalances.

Common parasites and their effects

External Parasites: Fleas, ticks, mites, and lice cause itching, hair loss, and skin infections.

Internal Parasites: Worms (roundworms, tapeworms, hookworms) can cause digestive issues, weight loss, and anaemia.

Cat with scabies, caused by a parasitic mite

Transmission of diseases, disorders, and parasites

Understanding how diseases spread helps in prevention and control. The key transmission methods include:

1. Direct transmission

Physical contact between animals (e.g., licking, biting, mating).

Example: Rabies is transmitted through bites.

2. Indirect transmission

Disease spreads via an intermediate object or organism.

Example: Parvovirus can survive on contaminated surfaces like food bowls.

3. Airborne transmission

Inhalation of pathogens present in the air.

Example: Canine distemper spreads through respiratory droplets.

4. Droplet transmission

Spread through moisture droplets from sneezing, coughing, or barking.

Example: Kennel cough is highly contagious via droplets.

5. Ingestion

Consuming contaminated food or water.

Example: Salmonella can spread through raw food.

6. Vector transmission

Disease-carrying organisms (e.g., mosquitoes, ticks) transmit infections.

Example: Lyme disease is spread by ticks.

7. Fomite transmission

Objects like bedding, clothing, and equipment act as disease carriers. These objects are called fomites.

Example: Ringworm spores can linger on grooming tools.

Impact on animals at different life stages

The severity and effects of diseases vary depending on an animal's age and role. Below is how different life stages are impacted:

Neonate (newborns)

High susceptibility due to underdeveloped immune systems.

Common issues: Neonatal septicaemia, hypothermia, dehydration.

Juvenile (young animals)

Developing immune responses but still vulnerable.

Common issues: Parvovirus, parasitic infections (roundworms, coccidia).

Adult animals (working, breeding)

Stronger immunity but still prone to stress-induced illnesses.

Common issues: Reproductive infections, respiratory diseases (kennel cough, equine influenza).

Senior animals (aging but still active)

Declining immunity and organ function.

Common issues: Arthritis, heart disease, kidney disease.

Geriatric animals (elderly and less active)

Highly susceptible to chronic illnesses.

Common issues: Cognitive dysfunction, cancer, diabetes.

> **Remember!**
> **Prevention is more effective than cure.**
> Preventing diseases through **vaccination, regular parasite control, hygiene, and biosecurity measures** is the most important aspect of animal management. Early detection and prompt treatment can save lives and reduce suffering.

Effects, prevention and treatment of diseases, disorders and parasites on animals

Understanding diseases, disorders and parasites is essential in animal management. These conditions can significantly impact the health, welfare, and productivity of animals, influencing their physiology, biological systems and successful rearing.

Effects of diseases, disorders, and parasites on animals

1. Health and welfare

- Diseases can cause pain, distress, and behavioural changes in animals.
- Some disorders, such as genetic conditions, may lead to chronic suffering.
- Parasites (e.g., fleas, ticks, worms) can cause irritation, weakness, and secondary infections.
- Reduced immune response due to prolonged illness can make animals more susceptible to further infections.

2. Physiology

- Infectious diseases can disrupt normal bodily functions, including respiration, digestion, and circulation.
- Parasites can drain essential nutrients, leading to deficiencies and metabolic imbalances.
- Some conditions (e.g., bacterial infections) produce toxins that affect organ function.
- Hormonal imbalances caused by disorders (e.g., diabetes) impact growth, reproduction, and metabolism.

3. Biological systems

- The immune system may be compromised, leading to reduced resistance against further infections.
- The cardiovascular system may be affected by parasites such as heartworms, leading to circulation problems.
- Respiratory infections (e.g., pneumonia) can decrease oxygen supply, impacting overall health.
- Digestive disorders can lead to malnutrition, weight loss, and poor coat condition.

4. Successful rearing

Infectious diseases can reduce reproductive success by affecting fertility and offspring viability.

Poor health in breeding animals may result in weaker litters and higher mortality rates.

Parasites can delay growth and development in young animals.

Chronic disorders can reduce an animal's value and productivity in commercial settings.

Prevention of diseases, disorders, and parasites

1. Personal Protective Equipment (PPE)

- Gloves, masks, and protective clothing help reduce the spread of infectious diseases.
- PPE is essential when handling sick animals, cleaning enclosures, and performing medical procedures.

2. Husbandry

- Proper nutrition supports a strong immune system.
- Regular exercise promotes healthy organ function and reduces stress.
- Clean and comfortable housing conditions prevent the spread of disease.

- Routine health checks ensure early detection of illnesses.

3. Biosecurity

- Quarantine measures for new or sick animals prevent disease transmission.
- Disinfection of equipment, surfaces, and enclosures minimises the risk of infection.
- Controlled access to animal areas limits exposure to potential pathogens.
- Vaccination programs protect against common diseases.

Treatment of diseases, disorders, and parasites

1. Health and welfare treatments

- **Supportive Care:** Includes fluid therapy, pain relief, and nutritional support to aid recovery.

- **Husbandry Adjustments**: Special diets, modified environments, and stress reduction techniques promote healing.

- **Biosecurity Measures**: Isolation of infected animals prevents disease spread and allows for targeted care.

2. Anatomy and physiology-based treatments

- **Medications**: Antibiotics for bacterial infections, antifungals for fungal infections, and antiparasitic drugs for internal and external parasites.

- **Surgical Interventions**: Tumour removal, fracture repair, and corrective procedures for congenital disorders.

- **Physiotherapy and Rehabilitation**: Supports recovery from musculoskeletal injuries or neurological conditions.

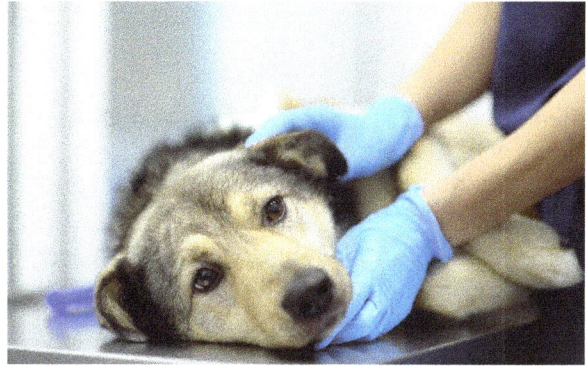

A dog being treated by a vet

Remember!
Remember! Early Detection and Prevention are the Best Strategies in Animal Health Management. Proactive prevention through biosecurity, proper husbandry, and vaccinations significantly reduces disease impact. Identifying early signs of illness ensures timely intervention, preventing suffering and economic loss in animal management.

Important terms!

Pathogen: A microorganism (bacteria, virus, fungus, parasite) that causes disease.

Fomite: An object or surface that carries infectious agents (e.g., bedding, equipment).

Vector: An organism (e.g., tick, mosquito) that transmits disease between animals.

Incubation Period: The time between infection and the onset of symptoms.

Endemic Disease: A disease that is consistently present in a certain population or area.

Notifiable Disease: A disease that must be reported to authorities due to its impact on public and animal health (e.g., Foot-and-Mouth Disease).

Parasitic Load: The number of parasites affecting an animal's body.

Immune Response: The body's defence mechanism against infections.

Biosecurity: Measures taken to prevent the introduction and spread of diseases.

Husbandry: The care, management, and feeding of animals to maintain health and productivity.

PPE (Personal Protective Equipment): Clothing and equipment used to protect individuals from exposure to hazards.

Immune System: The body's defence mechanism against infections and diseases.

Parasite: An organism that lives on or inside a host, deriving nutrients at the host's expense.

Vaccination: A preventive measure that stimulates the immune system to fight specific diseases.

Supportive Care: Treatment methods aimed at maintaining overall health and comfort during illness.

Quarantine: Isolation of animals to prevent the spread of contagious diseases.

Disinfection: The process of eliminating harmful pathogens from surfaces, equipment, and enclosures.

Notifiable diseases and required actions

Rabies

Overview: Rabies is a viral disease that affects the central nervous system in mammals, including humans. It is almost always fatal once symptoms appear.

Symptoms: Changes in behaviour, aggression, excessive drooling, paralysis and difficulty swallowing.

Actions if Suspected/Confirmed:

- Immediate notification to the Animal and Plant Health Agency (APHA).
- Quarantine measures to prevent spread.
- Euthanasia of infected animals in non-endemic areas.
- Vaccination programs for high-risk animals.
- Strict import regulations to prevent entry into the UK.

Avian influenza (bird flu)

Overview: A highly contagious viral infection affecting birds, with some strains capable of infecting humans.

Symptoms: Respiratory distress, swollen head, blue discolouration, diarrhoea, and sudden death.

Actions if Suspected/Confirmed:

- Notify APHA immediately.
- Restrict bird movements.
- Culling of infected and exposed birds.
- Disinfection of affected areas.
- Implementation of biosecurity measures, such as protective clothing for handlers.

Bovine spongiform encephalopathy (BSE) (mad cow disease)

Overview: A fatal neurodegenerative disease in cattle, linked to variant Creutzfeldt-Jakob disease (vCJD) in humans.

Symptoms: Nervousness, coordination loss, weight loss, aggression and difficulty walking.

Actions if Suspected/Confirmed:

- Notify APHA immediately.
- Culling and incineration of infected cattle.
- Ban on feeding cattle with animal-based protein to prevent spread.
- Surveillance and monitoring programs put in place.

Tuberculosis (TB) (bovine TB)

Overview: A bacterial infection affecting cattle and wildlife, transmissible to humans via unpasteurised dairy products.

Symptoms: Weight loss, cough, lethargy, swollen lymph nodes.

Actions if Suspected/Confirmed:

- Routine testing of cattle herds.
- Slaughter of infected animals.
- Movement restrictions on affected farms.
- Badger culling or vaccination in high-risk areas.

Bluetongue

Overview: A viral disease affecting ruminants (sheep, cattle, deer), spread by biting midges. It does not affect humans.

Symptoms: Swollen lips, face, tongue, fever, nasal discharge, lameness.

Actions if Suspected/Confirmed:

- Notify APHA immediately.
- Movement restrictions on infected farms.
- Vector control (midge population management).
- Vaccination programs in at-risk areas.

Foot and mouth disease (FMD)

Overview: A highly contagious viral disease affecting cloven-hoofed animals (cattle, sheep, pigs), with devastating economic impacts.

Symptoms: Fever, blisters in the mouth and on hooves, lameness, excessive drooling.

Actions if Suspected/Confirmed:

- Immediate reporting to APHA.
- Quarantine of affected premises.
- Culling of infected and at-risk animals.
- Restrict movement of livestock and vehicles.
- Disinfection of contaminated areas.

Newcastle disease

Overview: A viral disease affecting poultry and other birds, causing severe economic losses in the poultry industry.

Symptoms: Respiratory distress, nervous signs (twisted neck, tremors), green diarrhoea, sudden death.

Actions if Suspected/Confirmed:

- Notify APHA immediately.
- Culling of infected flocks.
- Movement restrictions and biosecurity enforcement.
- Vaccination in affected regions.

Zoonotic diseases

Understanding zoonotic diseases is crucial in animal management, as they can be transmitted between animals and humans. This guide covers key zoonotic diseases, their identification, actions taken when suspected or confirmed, and their implications for animal care professionals.

Ringworm

Cause: Fungal infection (dermatophytes)

Symptoms in animals: Hair loss, scaly skin, circular lesions

Symptoms in Humans: Red, itchy, ring-shaped rash

Action: Isolate infected animals, disinfect equipment, wear protective clothing, seek veterinary and medical advice, use antifungal treatment

Tapeworm

Cause: Parasitic flatworms (e.g., *Taenia*, *Echinococcus*)

Symptoms in Animals: Weight loss, lethargy, segments in faeces

Symptoms in humans: Digestive issues, weight loss, abdominal pain

Action: Regular deworming, proper disposal of animal faeces, avoid raw meat feeding, maintain hygiene

Rabies (notifiable disease)

Cause: Rabies virus (affecting nervous system)

Symptoms in animals: Aggression, excessive drooling, paralysis

Symptoms in Humans: Fever, confusion, paralysis, fatal if untreated

Action: Immediate veterinary/medical attention, euthanasia for infected animals, report to authorities, vaccinate animals in high-risk areas

Salmonella

Cause: Bacteria (*Salmonella* spp.)

Symptoms in Animals: Diarrhoea, vomiting, fever

Symptoms in Humans: Food poisoning, stomach cramps, fever

Action: Proper food handling, clean animal enclosures, hand hygiene, avoid cross-contamination in food preparation

Leptospirosis

Cause: Bacteria (*Leptospira* spp.)

Symptoms in animals: Fever, kidney/liver failure, jaundice

Symptoms in humans: Flu-like symptoms, liver/

Bats can be a source of zoonotic diseases such as rabies

kidney damage, severe cases can be fatal

Action: Vaccinate animals, prevent standing water contamination, wear protective gear when handling infected animals

Campylobacteriosis

Cause: Bacteria (*Campylobacter* spp.)

Symptoms in Animals: Mild diarrhoea or asymptomatic

Symptoms in Humans: Food poisoning, diarrhoea, fever

Action: Proper food hygiene, avoid raw pet food, handwashing after handling animals

Toxocara canis/cati (roundworms)

Cause: Parasites (*Toxocara* spp.)

Symptoms in Animals: Digestive disturbances, swollen belly, weight loss

Symptoms in Humans: Fever, organ damage (ocular toxocariasis can cause blindness)

Action: Regular deworming, faeces disposal, hand hygiene, prevent soil contamination

Toxoplasma gondii

Cause: Parasite (*Toxoplasma gondii*)

Symptoms in Animals: Usually asymptomatic, but can cause stillbirths in livestock

Symptoms in Humans: Flu-like symptoms, serious for pregnant women and immunocompromised individuals

Action: Avoid handling cat litter during pregnancy, wash hands thoroughly, cook meat properly

> **Important terms!**
>
> spp.: This stands for 'species plural'. It means that there are a number of different species covered by the genus. For example, *Toxocara* spp. covers all Toxocara species, including *Toxocara canis*, *Toxocara catis*, *Toxocaris leonina*.

Symptoms and transmission of common diseases

Effective disease management involves early detection, strict hygiene protocols and appropriate treatment plans to ensure animal welfare and public safety.

1. Bacterial diseases

Bacterial infections can spread rapidly and cause severe health issues in animals. Proper sanitation, vaccination, and antibiotic treatments (where applicable) are essential for control.

Leptospirosis

Symptoms: Fever, vomiting, muscle pain, jaundice, kidney failure, lethargy, loss of appetite, increased thirst, difficulty breathing, blood in urine.

Transmission: Contact with infected urine, contaminated water, soil, or through wounds on the skin. Rodents are common carriers.

Brucellosis

Symptoms: Fever, infertility, spontaneous abortion in pregnant animals, weight loss, swollen joints, lameness, depression, swollen lymph nodes.

Transmission: Direct contact with infected animals, ingestion of contaminated reproductive fluids, milk, or placenta. Can be zoonotic.

Tuberculosis

Symptoms: Chronic coughing, weight loss, respiratory distress, lethargy, fever, swollen lymph nodes, difficulty breathing.

Transmission: Airborne droplets, direct contact with infected animals, contaminated feed, inhalation of respiratory secretions.

Salmonella

Symptoms: Diarrhoea (sometimes bloody), fever, vomiting, dehydration, lethargy, anorexia, abdominal cramps.

Transmission: Ingestion of contaminated food, water, faeces, or via contaminated surfaces. Can infect humans through improper hygiene.

Campylobacter

Symptoms: Diarrhoea (can be watery or bloody), fever, abdominal pain, vomiting, weight loss, dehydration.

Transmission: Ingestion of contaminated raw or undercooked food, water, faeces, or contact with infected animals.

E. Coli

Symptoms: Diarrhoea, dehydration, abdominal pain, fever, weakness, kidney failure in severe cases.

Transmission: Ingestion of contaminated food, water, or faeces; direct contact with infected animals or environments.

Bordetella bronchiseptica

Symptoms: Harsh coughing, nasal discharge, fever, lethargy, respiratory distress, pneumonia in severe cases.

Transmission: Airborne droplets, direct contact with infected animals, contaminated surfaces (kennel cough in dogs and other animals).

2. Viral diseases

Viruses can be highly contagious and require strict management, including vaccination and biosecurity measures.

Rabies

Symptoms: Aggression, hypersalivation, difficulty swallowing, paralysis, confusion, behavioural changes, seizures, ultimately fatal.

Transmission: Bite from an infected animal, saliva entering wounds or mucous membranes.

Distemper

Symptoms: Fever, nasal discharge, coughing, neurological signs (seizures, twitching), vomiting, diarrhoea, thickened paw pads.

Transmission: Airborne droplets, direct contact with infected bodily fluids, contaminated surfaces.

Parvovirus

Symptoms: Severe vomiting, bloody diarrhoea, dehydration, lethargy, fever, anorexia, high mortality in puppies.

Transmission: Contact with infected faeces, contaminated environments, high resistance in the environment.

Influenza (canine, feline, equine, avian, swine)

Symptoms: Coughing, nasal discharge, fever, lethargy, loss of appetite, respiratory distress.

Transmission: Airborne droplets, direct contact, contaminated surfaces.

Feline herpes virus

Symptoms: Sneezing, conjunctivitis, nasal discharge, ulcers, eye infections.

Transmission: Direct contact, contaminated surfaces, aerosols.

Feline calicivirus

Symptoms: Ulcers in mouth, respiratory distress, fever, nasal discharge.

Transmission: Airborne droplets, direct contact with infected saliva or nasal secretions.

Feline leukaemia

Symptoms: Anaemia, weight loss, infections, swollen lymph nodes, lethargy.

Transmission: Saliva, nasal secretions, shared food bowls, blood transfusions.

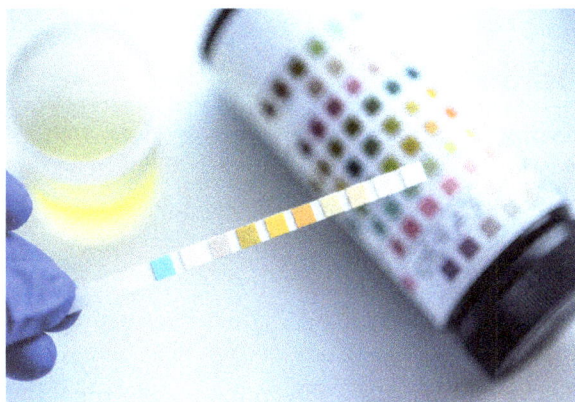

Testing urine for diseases

Feline immunodeficiency virus (FIV)

Symptoms: Weight loss, fever, infections, immune suppression, swollen lymph nodes.

Transmission: Bites from infected cats, blood contact.

Infectious canine hepatitis

Symptoms: Fever, vomiting, jaundice, abdominal pain, seizures in severe cases.

Transmission: Contact with urine, faeces, saliva.

Myxomatosis

Symptoms: Swelling around eyes, respiratory distress, lethargy, conjunctivitis, nodules on the skin.

Transmission: Fleas, mosquitoes, direct contact with infected animals.

Bluetongue

Symptoms: Swelling of tongue, fever, breathing difficulties, lameness.

Transmission: Biting insects (midges), direct contact.

Foot and mouth disease

Symptoms: Blisters on feet, mouth ulcers, fever, excessive salivation.

Transmission: Airborne, direct contact, contaminated equipment.

Rabbit haemorrhagic disease (RHD)

Symptoms: Internal bleeding, sudden death, fever, lethargy.

Transmission: Direct contact, insects, contaminated food.

3. Protozoan diseases

Giardia

Symptoms: Diarrhoea, weight loss, dehydration, weakness.

Transmission: Contaminated water or faeces.

Toxoplasma gondii

Symptoms: Neurological signs, fever, weight loss, reproductive issues.

Transmission: Ingesting contaminated meat or faeces.

4. Fungal diseases

Ringworm

Symptoms: Circular bald patches, flaky skin, itchiness, redness.

Transmission: Direct contact with infected animals, contaminated surfaces.

5. Prion diseases

Bovine spongiform encephalopathy (BSE)

Symptoms: Uncoordinated movements, nervousness, weight loss, fatal neurological degeneration.

Transmission: Ingestion of infected meat or contaminated feed.

Chronic wasting disease

Symptoms: Weight loss, behavioural changes, staggering, salivation.

Transmission: Direct contact, contaminated environments.

6. Parasites

Parasites can be divided into two categories: **endoparasites**, which live inside the host, and **ectoparasites**, which live on the skin or fur of the host.

Endoparasites

Roundworms (*Toxocara* spp.): Can cause weight loss, bloating, vomiting, and diarrhoea. Transmission occurs through ingestion of infected faeces, contaminated soil, or from mother to offspring.

Tapeworms (*Dipylidium caninum, Taenia* spp.): Can cause weight loss, poor coat condition, and segments visible in faeces. Transmission occurs through ingestion of infected fleas or raw meat.

Flukes (*Fasciola hepatica* – liver fluke): Primarily affects ruminants, causing liver damage, anaemia and weight loss. Transmission

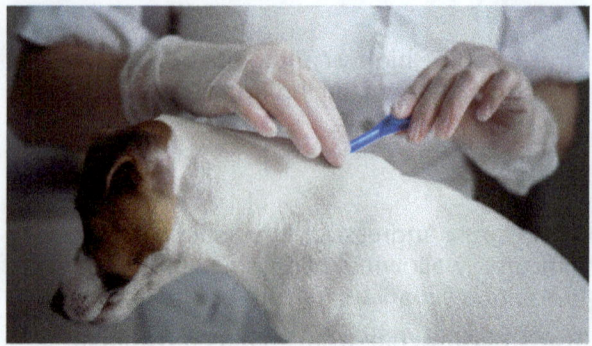

Treating a dog for parasites

> **Important terms!**
>
> Zoonotic Disease: A disease that can be transmitted between animals and humans.
>
> Pathogen: A microorganism that causes disease.
>
> Vector: An organism that transmits pathogens from one host to another.
>
> Endoparasite: A parasite that lives inside the body of its host.
>
> Ectoparasite: A parasite that lives on the surface of the host.
>
> Prion: An infectious protein that causes degenerative brain diseases.
>
> Vaccination: A method of stimulating immunity against a disease using an antigenic substance.
>
> Transmission: The method by which a disease spreads from one host to another.
>
> Incubation Period: The time between exposure to a pathogen and the onset of symptoms.

occurs through ingestion of contaminated vegetation or water.

Protozoa (*Coccidia, Giardia*): Cause diarrhoea, dehydration, and weight loss. Transmission occurs through contaminated water, faeces or soil.

Ectoparasites

Ticks (*Ixodes, Dermacentor* spp.): Can transmit Lyme disease and other pathogens. Found in tall grass and attach to the host's skin, feeding on blood.

Fleas (*Ctenocephalides* spp.): Cause itching, hair loss, skin irritation, and may transmit tapeworms. Spread through contact with infested environments or animals.

Lice (*Trichodectes canis, Linognathus setosus*): Cause itching, hair loss, anaemia, and skin infections. Transmission occurs through direct contact.

Mites (*Sarcoptes scabiei, Demodex* spp.): Cause mange, skin irritation and hair loss. Transmission occurs through direct contact with infected animals or surfaces.

Regular deworming, flea control, and maintaining clean environments are crucial for parasite prevention and management.

> ### Remember!
> A major concern in animal care is the **zoonotic potential** of many diseases, meaning they can transfer from animals to humans. Diseases such as **rabies, leptospirosis, salmonella, and toxoplasmosis** pose a direct threat to human health. Proper hygiene, biosecurity measures, and vaccination programs are essential to prevent outbreaks and protect both animals and humans.

Nutritional disorders and their causes/symptoms

Nutritional disorders arise from **imbalances in diet**, leading to deficiencies or excesses in essential nutrients. Below are the primary nutritional disorders encountered in animal care.

Rickets

Cause:
- Deficiency of **vitamin D, calcium, or phosphorus**.
- Common in young, growing animals.

Symptoms:
- Weak, soft bones prone to fractures.
- Bowed legs and skeletal deformities.
- Difficulty standing or moving.
- Poor growth rate.

Scurvy

Cause:
- Vitamin C deficiency.
- Common in species like guinea pigs, which cannot synthesise vitamin C naturally.

Symptoms:
- Swollen, painful joints.
- Lethargy and weakness.
- Poor coat condition.
- Delayed wound healing.

Anorexia

Cause:
- Underlying health conditions (e.g., infections, pain, dental disease).
- Stress or poor environmental conditions.
- Sudden dietary changes.

Symptoms:
- Loss of appetite.
- Weight loss.
- Weakness and lethargy.
- Dull coat and poor general health.

Obesity

Cause:
- Excessive food intake combined with **insufficient exercise**.
- High-calorie diet with poor nutrient balance.
- Genetic predisposition in certain breeds.

Symptoms:
- Excess body fat.
- Difficulty in movement or breathing.
- Increased risk of diabetes, arthritis, and heart disease.
- Lethargy.

Urolithiasis

Cause:
- Formation of **urinary stones** due to imbalanced minerals (e.g., excess calcium).
- Dehydration or poor water intake.
- Inappropriate diet.

Symptoms:
- Straining to urinate.
- Blood in urine.
- Pain and discomfort.
- Frequent urination with small amounts.

Laminitis

Cause:
- Excessive intake of **carbohydrates** (e.g., lush pasture, high-starch feeds).
- Obesity and metabolic disorders.
- Poor hoof care.

Symptoms:
- Lameness and reluctance to walk.
- Heat and pain in hooves.
- Abnormal hoof growth.
- Stiffness in movement.

Taurine deficiency

Cause:

- **Lack of taurine**, an essential amino acid, in diet.
- Common in **cats**, as they cannot synthesise taurine.

Symptoms:

- Vision impairment or blindness.
- Heart disease (**dilated cardiomyopathy**).
- Reproductive issues.
- Poor coat condition.

> **Remember!**
> Nutritional disorders are **preventable** with a balanced diet and appropriate husbandry. Proper feeding regimes and dietary management **reduce the risk of deficiencies and excesses**, ensuring optimal animal health.

Further disorders and their causes/symptoms

Understanding the various diseases, disorders, parasites, and notifiable diseases that can affect animals is crucial for professionals in animal care and management. This revision guide focuses on specific endocrine, physical, and metabolic disorders, detailing their causes, symptoms and implications.

Endocrine disorders

Endocrine disorders result from imbalances in hormone production, leading to various physiological issues.

- **Cushing's Disease (Hyperadrenocorticism):** This condition arises from excessive cortisol production by the adrenal glands. Causes include pituitary gland tumours (most common) or adrenal gland tumours. Symptoms encompass increased thirst and urination, excessive appetite, panting, a pot-bellied appearance, thinning skin, and hair loss.
- **Addison's Disease (Hypoadrenocorticism):** Caused by insufficient production of adrenal hormones (cortisol and aldosterone), often due to immune-mediated destruction of the adrenal cortex. Symptoms are vague and can include lethargy, vomiting, diarrhoea, weight loss and weakness. Severe cases may lead to an Addisonian crisis, characterised by shock and collapse.
- **Hypothyroidism:** This disorder results from inadequate production of thyroid hormones, commonly due to autoimmune thyroiditis or idiopathic thyroid atrophy. It is more prevalent in dogs than cats. Symptoms include lethargy, weight gain, cold intolerance, skin infections, and hair loss.
- **Hyperthyroidism:** Predominantly seen in older cats, hyperthyroidism is caused by overproduction of thyroid hormones, often due to benign thyroid adenomas. Symptoms include weight loss, increased appetite, hyperactivity, vomiting, diarrhoea, and increased thirst and urination.

Physical disorders

Physical disorders pertain to structural or functional impairments affecting an animal's mobility or general health.

- **Lameness:** A clinical sign indicating pain or dysfunction in the limbs, lameness can result from various conditions, including injuries, arthritis, or infections. Symptoms include limping, reluctance to move and abnormal gait.
- **Egg Binding:** This condition occurs when a female bird is unable to pass an egg through the reproductive tract. Causes include nutritional deficiencies (e.g., calcium), obesity, lack of exercise, or oversized eggs. Symptoms encompass abdominal straining, lethargy, fluffed feathers and laboured breathing.
- **Arthritis:** A degenerative joint disease leading to inflammation and pain in the joints. Causes include aging, joint instability, or previous injuries. Symptoms are stiffness, limping, reduced activity and discomfort when touched.

Metabolic disorders

Metabolic disorders involve disruptions in the body's chemical processes, affecting energy production and utilisation.

- **Diabetes Mellitus:** Characterised by insufficient insulin production or action, leading to elevated blood glucose levels. Causes include genetic predisposition, obesity and pancreatitis. Symptoms involve increased thirst and urination, weight loss despite increased appetite, lethargy, and cataracts in dogs.

- **Metabolic Bone Disease (MBD):** Commonly seen in reptiles, MBD results from calcium deficiency or improper calcium-phosphorus balance, often due to inadequate diet or lack of UVB lighting. Symptoms include soft or deformed bones, fractures, lethargy and muscle tremors.

- **Milk Fever (Hypocalcaemia) / Eclampsia:** Typically affecting lactating animals, this condition arises from low blood calcium levels. Causes include inadequate dietary calcium or excessive calcium loss during milk production. Symptoms are muscle tremors, restlessness, fever and, in severe cases, seizures.

Recap Questions

1. What are the seven main methods of disease transmission in animals?

2. How can diseases and parasites affect animals differently at various life stages?

3. What are some common signs and symptoms of respiratory diseases in animals?

4. What are three ways diseases, disorders and parasites can affect the biological systems of animals?

5. What are three key biosecurity measures used to prevent the spread of diseases in animal care?

6. What is the primary goal of supportive care in disease treatment, and what are two examples of it?

7. What immediate action must be taken if a notifiable disease is suspected or confirmed in animals?

8. Which disease mentioned in the text is spread by a vector, and what organism transmits it?

9. Why is culling used as a control measure for some animal diseases, and provide one example from the text where it is implemented?

10. What actions should be taken when an animal is suspected of having ringworm?

11. Why is rabies classified as a notifiable disease, and what immediate actions must be taken if an animal is infected?

12. How can the spread of zoonotic diseases like salmonella and campylobacteriosis be prevented?

13. What is a zoonotic disease, and can you name two examples from the text?

14. How is leptospirosis transmitted, and what are common symptoms of the disease?

15. What is the difference between an endoparasite and an ectoparasite? Provide one example of each.

16. What is a common cause of rickets in young, growing animals?

17. Why is taurine essential in a cat's diet, and what condition can its deficiency cause?

18. What are two potential causes of obesity in animals?

19. Which endocrine disorder is caused by excessive cortisol production and can result in symptoms such as increased thirst, excessive appetite and hair loss?

20. What metabolic disorder in reptiles is linked to calcium deficiency and lack of UVB lighting, leading to symptoms such as soft bones and muscle tremors?

Important terms!

Nutritional Disorder: A health condition caused by improper diet, either through deficiency or excess of nutrients.

Deficiency: A lack of essential nutrients needed for normal body functions.

Vitamin D: A nutrient required for calcium absorption and bone health.

Calcium: A mineral essential for strong bones and teeth.

Taurine: An amino acid crucial for heart and eye health in cats.

Metabolic Disorder: A condition affecting the body's ability to process nutrients.

Lameness: Difficulty in movement due to pain or weakness in limbs.

Cardiomyopathy: A disease affecting heart muscle function.

Adrenal Glands: Small glands located above the kidneys that produce hormones like cortisol and aldosterone.

Cortisol: A steroid hormone involved in stress response, metabolism, and immune function.

Thyroid Gland: An endocrine gland in the neck producing hormones that regulate metabolism.

Insulin: A hormone produced by the pancreas that regulates blood glucose levels.

UVB Lighting: Ultraviolet B light necessary for vitamin D synthesis, crucial for calcium metabolism in reptiles.

Polydipsia: Excessive thirst.

Polyuria: Excessive urination.

Polyphagia: Increased appetite.

Lethargy: A state of sluggishness, inactivity, and apathy.

Metastasis: The spread of cancer cells from the original site to other parts of the body.

Pancreatitis: Inflammation of the pancreas.

Seizures: Uncontrolled electrical activity in the brain, leading to convulsions or other physical manifestations.

Tachycardia: Abnormally rapid heart rate.

Real world example: A sheep farm

Imagine you're on a sheep farm. The sheep are raised for their wool and meat, and everything seems to be running smoothly. However, one day, some of the sheep start acting differently — they're lethargic, have a loss of appetite, and appear to be losing weight rapidly.

Diseases and Disorders: The farmer notices that some of the sheep have **foot rot**, a common **disorder** that affects the hooves of sheep. It causes pain, swelling and limping. Foot rot is caused by a bacterial infection and, if untreated, can make it difficult for sheep to walk and graze, which impacts their health and the farmer's productivity.

Another problem that arises is a disease called **Bluetongue**. This is a viral disease spread by biting midges, and it causes fever, swelling of the tongue (which turns blue), and lameness. It's particularly harmful to sheep and cattle and can sometimes lead to death. In this case, the farmer notices that some sheep have swollen faces and are having trouble eating and drinking.

Parasites: The farm also has a **worm** problem, especially **liver flukes**. These parasites live in the liver of infected animals. The sheep start showing signs of **weight loss, anaemia** and **jaundice**. Liver fluke infections are a concern because they can cause severe damage to the liver and, if left untreated, may lead to death. The worms are often carried by snails in the environment, and the sheep can become infected by grazing on contaminated grass.

Notifiable Diseases: In addition to these, the farm has to be particularly careful about **notifiable diseases**, which are diseases that must be reported to authorities. One such notifiable disease is **Foot-and-Mouth Disease**. If one of the sheep on the farm is suspected to have Foot-and-Mouth Disease (FMD), it is required by law to notify local veterinary authorities. FMD is highly contagious and can spread rapidly among cattle, sheep and pigs. It causes blisters in the mouth and hooves, making it extremely painful for animals and affecting their ability to eat or walk. This disease has major implications for agriculture because it can cause trade restrictions and economic loss.

What Happens Next? The farmer calls the vet, who confirms that some sheep have foot rot and treats them with antibiotics and anti-inflammatory drugs. The vet also checks for Bluetongue and advises the farmer to use insect repellents to prevent further spread of the disease. The vet treats the liver fluke infection with a dewormer and recommends managing pasture conditions to prevent snails from spreading the parasite.

Finally, the vet reminds the farmer to stay alert for signs of Foot-and-Mouth Disease, and if there is any suspicion of it the farmer must immediately report it to local authorities, to prevent an outbreak.

7.3 First Aid for animals including for wounds and conditions using a first aid kit

First aid is the immediate help and care given to an animal that is acutely injured or ill. It is given by the person who is with the animal at that time.

Aims and limitations

Aims

The aims of first aid are:

- Preserve life
- Prevent the situation worsening
- Promote recovery

They are often referred to as the 3Ps of first aid.

Limitations

First Aid is very limited other than the 3Ps.

It:

- Does not always diagnose the problem.
- Gives time for the animal to get to a vet.
- Can be life-saving.

Considerations relating to legislation

- An animal's welfare must always be considered in a first aid situation.
- Under the Animal Welfare Act 2006 there is a duty of care to take reasonable steps to ensure that the animal's needs are met by the person responsible for the animal. This includes its need to be protected from pain, suffering, injury and disease. Therefore, first aid is part of the duty of care towards an animal.
- The Veterinary Surgeons Act 1966 details what can and can't be carried out by qualified and non-qualified persons (detailed in the previous section). First aid is allowed by non-qualified persons under this act. However, it is essential that the animal is taken to the vets for checking and further treatment by a qualified person.

Hazards

A hazard is something in the environment that poses a risk to either the animal or the person carrying out the first aid.

Assessing the environment for hazards and reducing risk to humans and animals

- When confronted with a first aid situation the environment should be assessed first before acting.
- Hazards can include vehicles, sharp objects, other animals, water, fire, poisons, chemicals, electrical wiring and temperature extremes.
- It might be necessary to move the animal sooner than ideal to move it out of a hazardous area such as near high drops, close to water, out of water or out of fire / smoke filled spaces. This is in order to prevent the situation from worsening.
- It is important not to make the situation worse by acting when the hazards are too great.
- For example:
  » A cat that has had a road traffic accident may be at risk of being hit by another car or the person carrying out the first aid may be at risk of being hit by a car.

  » Dogs injured in a dog fight may continue to fight whilst first aid is attempted leading to further injuries for the dogs or the first aiders.

  » The hazards in the environment may be the injured/ill animal themselves as an injured or ill animal may be aggressive towards the first aider.

- It is essential to take time to be calm and assess the situation before acting in first aid situations. The question should be asked if it is safe to approach the animal.
- When dealing with the animal the first aider should be calm and reassuring with no sudden movements. This will reduce the risk of frightening the animal and causing further problems.
- The animal should be handled and moved as little as possible depending on the first aid required.

First aid kit contents and purposes

The recommended contents of an animal first aid kit with their purposes are as follows:

Item and Purpose	Image
Selection of bandages • Variety of sizes needed. • Cover wounds once dressed. • Help stop bleeding.	
Cotton wool • For cleaning injuries. • Padding for wounds to absorb blood.	
Sterile dressing materials • To be used directly onto site of injury (not burns) before using bandages etc.	
Adhesive tape • To secure bandages and dressing materials in place.	
Rectal thermometer • Not commonly kept in first aid kits. • Can be used to check body temperature. • Need to be careful to take temperature of rectal wall rather than faeces.	
Tweezers • Removal of foreign bodies such as splinters, thorns grass sides and stings.	
Gloves • To protect animal and humans from infections when dealing with injuries. • Need to ensure that they do not trigger allergies in animals or humans.	
Scissors • For cutting bandages and dressings and for trimming hair/fur away from edges of wound if necessary. • Blunt-nosed scissors are best to avoid additional stabbing injuries.	

Item and Purpose	Image
Hand sanitiser • To clean hands before dealing with any injuries to limit animal infections and also afterwards to prevent human infections.	
Eye wash • Not commonly kept in animal first aid kits. • Can be used to wash foreign bodies out of the eye such as liquids and grass seeds. • Eyes should be checked by a vet.	
Antiseptic solution • To reduce infection in wounds once cleaned and to kill pathogens. • There are antiseptic sprays that kill 99.9% of bacteria, virus and fungal spores on contact.	
Tick remover • To safely remove ticks from the skin including the mouthparts. • Full removal of the tick reduces chances of secondary infection. • Ticks can be infected with and transmit other pathogens, for example the Lyme disease bacteria.	
Carrier bag • To take away any waste materials after first aid so can be disposed of suitably. • To help transport animal to vets if small enough for bag.	
Blanket • To keep an animal warm. • To be used as a makeshift stretcher.	

The contact details of the local veterinary practice should be kept with the first aid kit along with details for out of hours, emergency service.

Applying first aid treatment

Wounds

In all situations the vet should be contacted if wound symptoms are severe or require further advice. Remember first aid is there to give time to get the animal to veterinary treatment.

Wound Type	First Aid to apply
Abrasions	• Caused by a scraping force or friction. • Clean wound to clear any debris and to reduce infection. • Apply antiseptic solution. • Cover wound to protect it and stop any bleeding. • Bandage if necessary (not too tightly). • Seek veterinary advice if bleeding does not stop.
Lacerations	• This is a torn, jagged wound caused by a blunt trauma or tearing force (such as another animal biting). • Less bleeding however more damage to surrounding tissues. • Clean wound to clear any debris and to reduce infection. • Apply antiseptic solution. • Cover wound to protect it and stop any bleeding. • Bandage if necessary (not too tightly). • Seek veterinary advice if bleeding does not stop.
Concussions	• This is a type of brain injury that is caused by a blow or jolt to the head. • Can also be due to an injury to another part of the body that transmits force to the head. • Symptoms vary but include: lethargy, unresponsiveness, disorientation, difficulty walking, vomiting, seizures or unconsciousness. • Animal should be taken to vets immediately whilst monitoring breathing and respiration and kept warm and reassured.
Contusions	• These are bruising to the skin, where the tissue is damaged underneath. • Aim is to prevent further damage. • A cold compress or ice wrapped in a towel can be applied to the affected area.
Punctures	• These wounds can be caused by bites (teeth), stepping on a sharp object or being struck by a sharp object (such as a stick). • Clean wound to clear any debris and to reduce infection. • Apply antiseptic solution. • Cover wound to protect it and stop any bleeding. • Bandage if necessary (not too tightly). • Seek veterinary advice if bleeding does not stop.
Incisions	• This is a straight cut made by a sharp object such as a knife. • Tend to be associated with surgery and dealt with as below for surgical wounds.
Surgical wounds	• Surgical wound should have been dealt with at the vets. • They may need redressing, and instructions would have been supplied by vet. • If the wound becomes red, hot and there is pus then the wound can be cleaned and antiseptic applied, followed by check up with the vet as medication may be required or wound retreated.

Wound Type	First Aid to apply
Thermal wounds	• These are burns (dry heat) and scalds (wet heat). • The severity of the wound can vary depending on depth of tissue affected. • The affected area should be cooled with water. • No creams or ointments should be applied or blisters broken. • Keep the animal warm and quiet. • Bandages may stick to the wound so clingfilm can be used to wrap the wound to keep it clean whilst transporting to the vets.
Chemical wounds	• Can be from wet or dry chemicals such as detergents. • Dry chemicals can be brushed off the animal. • Wet chemicals should be flushed with water and if in the eye with water or saline solution if available. • Vet should be contacted to discuss next steps.
Electrical wounds	• Electrical injuries can cause dry heat burns (as above). • There can also be internal injuries so any animal should be checked with a vet and monitored carefully for breathing and shock
Bites	• Insect bites can cause allergic reactions or become infected,. In these cases, veterinary treatment will be required. • If the bite is close to the face, it can cause swelling of the airways and breathing needs to be monitored as the animal is taken to the vets. • Bites from other animals can cause lacerations (see above).
Stings	• Removal of the sting if possible is recommended, carefully ensuring it is not broken and part of it left in the wound. • The swollen area can be treated with a cold compress or wrapped ice. • The animal should be discouraged from itching the area to avoid making it worse. • If the sting is on the face, it can cause swelling of the airways and breathing needs to be monitored as the animal is taken to the vets. • Allergic reactions can occur which can be severe, so the animal should be monitored carefully and if there are any concerns, a vet should be consulted.
Fractures	• A fracture is a break or crack in the bone. • All suspected fractures should be seen by a vet. • More complex fractures are those where the bone breaks through the skin or causes internal bleeding. • Try to stop the animal from moving the affected bones, splints can be used. • Cover any open wounds where possible with dressings and bandages to help control any bleeding and prevent infection.
Foreign bodies	• Foreign bodies are items that are not expected to be embedded in the animal's tissues (skin or eye). • Prevent the animal from touching the area and restrict their movements. • Do not remove the item as it may be stopping further bleeding. • Bleeding should be controlled where possible without pushing the object further into the wound. • No attempt should be made to remove anything embedded in the eye. • A vet should be seen immediately, keeping the animal warm and monitoring breathing.

Wound Type	First Aid to apply
Haemorrhage	• This is heavy bleeding, and an animal can deteriorate quickly. • It may be obvious external bleeding due to a wound in which case control the bleeding by using dressings and bandages and applying pressure (if no foreign objects). • If the bleeding is internal it may not be obvious. However, the animal may have pale gums, a rapid pulse or breath and a slow capillary refill time (CRT). • In both cases the animal should be kept warm and quiet due to shock and a vet consulted immediately.
Swelling	• Swelling can occur for a variety of reasons such as fractures, sprains, contusions, bites and stings. • The cause of the swelling should be established if possible. • A cold compress or wrapped ice could be applied and the animal rested. • Monitoring should be carried out for any signs of infection or deterioration. • Vet contacted depending on cause, severity and impact on animal welfare.

Conditions

Condition Type	First Aid to apply
Shock	• This is a life-threatening condition. • There is a lack of blood supply to the major organs possibly including the brain. • There are various stages of shock with the most severe being usually irreversible and fatal. • Aim of first aid is to stop the stages progressing by keeping the animal quiet and warm, monitoring breathing, controlling any bleeding, lowering the head to help blood flow to the brain. • Take the animal to the vets.
Convulsions	• These are also known as seizures and fits, where the animal can have uncontrollable movements or even be quite still. • There can be a variety of causes including illness and temperature. • The longer the convulsion lasts the more concerning it is, and a vet should be seen after one that lasts for a few minutes or if there are concerns after a short one. • Aim is to keep the animal safe, warm and quiet during the seizure and after and limit the lighting brightness. • Do not attempt to hold the animal or given them food or water.
Respiratory arrest	• If the animal is not breathing, then whilst the vet is called the animal should be laid on its side. • Its tongue should be pulled to one side if possible. • Airways should be checked for blockages. • The mouth should be closed and air blown into the nostrils until the chest rises and repeated till the animal breathes on its own or the vet is present. • There should also be a check to see if there is a heartbeat and if not place one hand under the chest for support and place the other over the heart (just behind the left front elbow). Then press down 100-120 times per minute using force appropriate to the size of the animal. • Compressions can be alternated with rescue breaths (around 30 compressions with two rescue breaths). • This should be continued until there is a heartbeat or the vet is present.

Condition Type	First Aid to apply
Choking	• Check to see if anything is visibly blocking the airway and can be removed without pushing it further in or being bitten. • Attempts can be made to dislodge the item by squeezing the ribcage or striking the ribcage. • Consult a vet if the item cannot be dislodged.
Poison	• The animal should be taken to the vets. • Do not attempt to make the animal vomit unless asked to by the vet. • If you know what the animal has consumed take it with you to the vet.
Allergens	• Allergens can cause allergic reactions which may need first aid. • Cooling the area affected can help. • If the reaction is around the head and neck area it can cause swelling, which affects breathing if severe (anaphylaxis). • Consult a vet.
Fly strike	• This occurs when faeces become impacted around the anus and flies lay eggs which hatch out into maggots. • You can try to gently soak the area to remove the faeces and maggots and trim the fur/feathers. • If it is severe or covers a large area veterinary treatment will be needed.
Hyperthermia	• When an animals temperature increases to a dangerously high level. • Cool animal using cool water or a fan, do not use ice, till panting reduces whilst keeping it safe and consult a vet . • Do not pour water over the head in case it leads to drowning.
Hypothermia	• When an animals temperature decreases to a dangerously low level. • Keep the animal warm and safe and consult a vet, hot water bottles, blankets and a warm fan or hair drier can be used to warm up the animal.
Electrocution	• Animals can become electrocuted due to chewing electrical wires. • Turn off the power before attempting to touch the animal. • Avoid water that has been in contact with an electrical source. • Check breathing and pulse (act accordingly). • Examine the animal for burns (act accordingly for a dry burn). • Keep the animal warm and quiet and consult the vets.
Severe dehydration	• Provide small amounts of water every few minutes or ice to lick. • Keep them cool and safe. • Consult a vet.

Seeking professional advice

If in any doubt what to do in a first aid situation, a call should be made to a veterinary practice or emergency out of hours service who will be able to provide advice and instructions over the phone.

Animals should be monitored carefully after first aid treatment and if not already being taken to a veterinary practice, then should be if there is any deterioration in condition or worrying symptoms.

Considerations when applying first aid

Do not forget:

• 3Ps are aims of first aid

• It is not a diagnosis

• Restrictions on what a non-qualified person can do

• Animal Welfare Act 2006 and the duty of care to protect animals from pain, suffering, injury and disease

• Assessing hazards in the environment

Further considerations to provide effective first aid

Weather. This can impact on the first aid required if there are weather extremes such as heat or cold. Wind, rain and snow can impact on the ability to carry out first aid effectively and also interfere with materials being used.

Location of wound. If the wound is not easily accessible it can limit what can be done in a first aid situation. It is best to get the animal to a veterinary practice as quickly as possible in these situations. It may be hard to bandage a wound depending on its location and whether there is a large protruding foreign object or bone. Large foreign objects are often stopping blood loss and should not be removed by a first aider. Bandaging should be applied to not put pressure on the object and cause further damage.

Husbandry conditions. If the animal is part of a group, it may be necessary to remove it from the group either by moving the animal elsewhere or by removing the rest of the group. This can be for the safety of the animal requiring first aid and for the first aider. It may be that the environmental conditions are such that the animal should be removed to a cleaner site once it is established that the animal is breathing and can be moved to prevent infection before further first aid is completed. It could also be that the lighting, ventilation, temperature or humidity of the environment require the animal to be moved as they are impacting in the first aid for animal and/or human.

Restraints. Animals can become fearfully aggressive or aggressive due to pain even if they are familiar with the first aider. It is wise to be hesitant and where necessary use restraints and PPE such as leads, muzzles, graspers, gloves, gauntlets nets, crush cages and towels to safely constrain the animal so it can be treated for first aid without it injuring itself further or the first aider. Care should be taken if lifting larger animals to ensure safe lifting practice to avoid human injuries.

Type of medication application. In first aid situations it is unlikely that any enteral or parenteral or inhalation routes will be used to administer medication. Remember first aid is all about the 3Ps so that the animal can then be treated by a qualified person such as a vet. It is possible that topical medication such as antiseptic sprays or creams may be used, for example on wounds. However, it could be that the animal has prescribed medication that the owner is allowed to administer, such as asthma inhalers or EpiPens. In this case, a first aid situation might require this medication to be given.

Remember!

First aid is not a diagnosis however it can be lifesaving.

The 3Ps of First Aid aim to give time to get the animal to a veterinary surgeon.

Hazards in the environment for humans and animals must be assessed before embarking on first aid.

Important terms!

3Ps of first aid: Preserve life, prevent the situation worsening, promote recovery

Hazards: Something in the environment that poses a risk

Wounds: Different types of injuries that can require first aid. Sixteen are listed in the syllabus

Conditions: Different types of illnesses that can require first aid. Eleven are listed in the syllabus.

First aid kit: Specific pack of items that can be used in a first aid situation to help achieve the three aims of first aid.

Recap Questions

1. What are the 3Ps of first aid?

2. Why is first aid limited?

3. Which of the five needs from the Animal Welfare Act 2006 relates to first aid?

4. Why are hazards important to considered before carrying out first aid?

5. Which type of scissors are best to include in a first aid kit?

6. What could you include in a first aid kit to help with ticks?

7. What types of thermal wounds are there?

8. What is a contusion and how can it be treated?

9. Why is an insect sting on the head or neck area potentially life-threatening?

10. Should foreign bodies be removed from the animal? Explain your answer.

11. What is the difference between hyperthermia and hypothermia?

12. What hazard should be dealt with before first aid treatment in electrocution?

13. Which medicines might be used in a first aid situation?

Practice Questions

1. Name two laws that govern the use of animal medication and give two examples each of how they meet their requirements. (2 marks)

2. Describe what a POM-VPS medicine is and give an example of a type of medication that is in this category. (3 marks)

3. Describe a scenario where a named wild animal may require medication. (3 marks)

4. Explain what anti-emetic and emetic medicines are and situations in which they might be used. (4 marks)

5. State the law which allows the use of controlled drugs in veterinary practices? (1 mark)

6. Controlled drugs cabinets must adhere to the Safe Custody Regulations 1973. List three features of design or construction that are required by these regulations. (3 marks)

7. There are four SQP categories relating to the species trained upon. Name two of the categories. (2 marks)

8. Describe why stock control is important for veterinary medicines, giving two examples. (4 marks)

9. Explain the difference between parental and enteral routes of medicine administration. (4 marks)

10. Explain why Schedule 3 of the Veterinary Surgeons Act 1966 is important in the medical treatment of animals. (4 marks)

11. List three key things a veterinary surgeon must consider before delegating a task to a registered vet nurse. (3 marks)

12. List three notifiable diseases that can affect livestock in the UK. (3 marks)

13. Explain the difference between a disease and a disorder in animals, providing an example of each. (4 marks)

14. A farmer notices that several cows in their herd are displaying signs of a parasitic infection. Describe the steps they should take to diagnose and manage the issue. (5 marks)

15. Compare and contrast bacterial and viral diseases in animals, focusing on their transmission, symptoms, and treatment options. (6 marks)

16. A vet is called to investigate an outbreak of a notifiable disease on a farm. Assess the potential consequences for the farmer, the local community, and animal welfare, and suggest measures to mitigate the impact. (8 marks)

17. Explain how internal and external parasites affect animal health, including examples of each and their impact on productivity. (5 marks)

18. A pet owner reports that their dog has been scratching excessively and has developed red, irritated skin. Using your knowledge of animal disorders and parasites, suggest possible causes and a course of action. (5 marks)

19. Analyse the potential risk factors for the spread of zoonotic diseases in a mixed-animal farm setting and propose strategies to minimise the risk of transmission to humans. (6 marks)

20. State the three aims of first aid. (3 marks)

21. Describe three situations where hazards in the environment may limit what first aid can be offered immediately. (6 marks)

22. List three items that are commonly included in a first aid kit and describe how they are used. (6 marks)

23. Name two types of wounds and describe the first aid that should be carried out for each. (6 marks)

Tackling the 12 mark questions

How to award marks

Marking 12-mark questions can be difficult. There are lots of correct statements that a learner can make, but how well does it answer the question?

- The key point to remember is that making 12 points does **NOT** get you 12 marks.

- The questions are marked in bands rather than the number of points made.

If you look at the sample mark schemes provided by the exam board, you will see that there is something called **indicative content**. This is a list of all the content that the awarding body believes you should be considering in your answer. They will look at this list and your answer, to decide which band your answer falls into. Then they will decide how many marks in that band your answer is worth.

The 12-mark questions may be only based on one unit, or they may cover several units, so ensure that you:

- identify which units relate to the questions

- have covered these units.

There will **ALWAYS** be **TWO** parts to a 12-mark question, so ensure you answer **BOTH** parts.

Let's look at the following example 12 mark question:

"A wildlife rehabilitation centre that cares for both wild and domestic rabbits has confirmed a case of the notifiable disease, rabbit haemorrhagic disease (RHD). The infected rabbit was housed in an outdoor enclosure near other rabbit enclosures.

Analyse the implications of an outbreak of rabbit haemorrhagic disease on the rehabilitation centre. Justify the control measures needed to prevent the spread of the disease and ensure the health and welfare of all other rabbits in the facility."

Now let's look at those bands for marks:

Basic	Good	Thorough	Comprehensive
1-3 marks	4-6 marks	7-9 marks	10-12 marks

> **Remember!**
> Remember! You are awarded marks based on literacy skills as well as demonstrating a breadth of knowledge.

This means that, for the above question, if you demonstrate a basic understanding of what RHD is, and the control measures that should be put in place, the most marks you will receive is 3. Whereas a thorough understanding, which explains how this virus spreads, the control measures and the implications of this, then you can receive up to 9 marks.

You must always remember to identify the command words in the question and understand what they want you to do. In this question you must ensure that you have **analysed** the implications and **justified** your answer - these are the two parts to the question. If you do not, then you will limit the band your answer is put in, and hence limit the number of marks.

For example, you may have given a really descriptive answer that covers lots of things. However, if you have not analysed and justified your descriptions, then your answer may only fall into the basic band.

See if you can match the exemplar answers on the next page to each of the categories. Look at the different approach of each answer and award an overall mark. Can you explain what makes one answer better than the other?

Exemplar answers	Marks
A highly contagious and frequently fatal virus, rabbit haemorrhagic disease (RHD) affects rabbits. Serious repercussions could result from an outbreak at the wildlife rehabilitation centre, such as high mortality rates, quick infection spread to other rabbits, and possible harm to the centre's finances and reputation. The afflicted rabbit should be isolated right away, and any other rabbits that may have come into contact with it should be closely watched in order to stop the spread of RHD. To stop indirect transmission, strict biosecurity protocols should be followed, which should include thoroughly disinfecting staff uniforms, equipment, and enclosures. To lessen the chance of further spread, the centre should also limit visitor access to rabbit enclosures. Vaccination of all remaining healthy rabbits would be an essential step in protecting them from infection. Additionally, staff should implement good hygiene practices, such as washing hands and changing clothing before handling different groups of rabbits. These measures will help reduce the risk of further cases and ensure the welfare of the rabbits in the centre. However, further steps, such as notifying relevant authorities and reviewing long-term biosecurity protocols, may be needed to fully control the outbreak.	
A highly contagious virus that affects rabbits, rabbit haemorrhagic disease (RHD) can spread quickly in a rehab centre. An outbreak might affect the centre's capacity to care for other animals and result in the death of infected rabbits. The centre should isolate infected rabbits and clean enclosures to stop the spread. Employees should limit their interactions with the rabbits and wear protective clothes. By taking these steps, the chance of developing new infections is decreased.	
Rabbit haemorrhagic disease (RHD) is a very contagious and typically fatal viral disease in rabbits. It is spread through direct contact with infected rabbits, infected items (e.g., food, bedding, or equipment), and even insects such as flies. Because of its high transmission rate and mortality rate, an outbreak at the wildlife rehabilitation centre has serious consequences for both the rabbits and the facility's operations overall. **Effects of an RHD Outbreak to the Rehabilitation Centre** One of the most significant impacts is the threat to the rabbit population, as the disease has a high death rate and typically results in sudden death with minimal warning. This would likely result in a significant loss of animals, especially if the disease were not quickly contained. Furthermore, as RHD is a notifiable condition, the centre would be under legal obligation to notify the concerned authorities of the outbreak, and this can lead to mandatory measures like quarantines or culling of diseased rabbits. The outbreak would also impact the reputation and visitor numbers of the centre. If the public is aware of the disease, biosecurity and animal welfare issues could lead to reduced funding or fewer visitors, at least temporarily when the centre must close to contain the spread. The cost of the outbreak would also be considerable, with extra expenses of veterinary treatment, disinfection, and biosecurity measures. **Justification of Control Measures to Prevent the Spread** To control the outbreak and conserve the remaining rabbits, infected and exposed rabbits must be separated as soon as possible. All affected rabbits, which include those showing signs of lethargy, fever, or bleeding, must be quarantined in a safe, separate room. Stricter biosecurity measures need to be practiced. The employees need to use protective clothing, gloves, and shoe covers and adhere to strict disinfection methods when moving between cages. Equipment, food trays, and cages need to be disinfected and cleaned with an accepted virucidal compound. Contaminated materials and bedding should be appropriately disposed of in order not to spread further. Restricting the movement of human and animal traffic within the facility is essential. Visitor numbers should be limited, and staff who come into contact with infected rabbits should not touch healthy ones. In addition, vaccination schemes for susceptible rabbits should be initiated where feasible, since RHD vaccinations can diminish the threat of infection considerably. Environmental management, such as the reduction of insect vectors (fleas and flies) that carry the virus, should also be considered. This could involve routine waste disposal, as well as ensuring enclosures are clean and secure. **Conclusion** An RHD outbreak poses a significant risk to the rabbits in the rehabilitation centre, as well as to the business's financial and reputational interests. Immediate and effective biosecurity, quarantine, and disinfection measures must be implemented to prevent further transmission and ensure the health and welfare of all remaining animals. Vaccination and ongoing disease surveillance are also required for long-term prevention.	

Planning your 12-mark answer

On average, learners who take the time to write a plan before answering a 12-mark question achieve a higher mark than those who don't. Writing a plan helps you to note down all the points that you want to make and then structure your answer in a sensible way.

Remember! If you run out of time an exam marker may award some marks based on your plan alone, so do not cross it out unless you are sure you have covered all of its content in your answer!

Let's look at the following example 12 mark question:

"An adult cat at an animal shelter has been diagnosed with feline calicivirus (FCV), a viral infection that causes respiratory issues. The cat is sneezing, has nasal discharge, and is showing signs of oral ulcers, as well as a reduced appetite. The cat is receiving supportive care and is currently isolated from other animals.

Analyse how feline calicivirus affects the respiratory and oral systems of the cat. Justify the steps the shelter needs to take to ensure the health of the cat during its isolation and recovery process. You are not expected to discuss any specific medications administered."

Have a go at planning out how you would answer this question.

Remember to consider the command words the question is using.

There are two templates provided. Try both and see which you prefer.

Template 1: Mind map

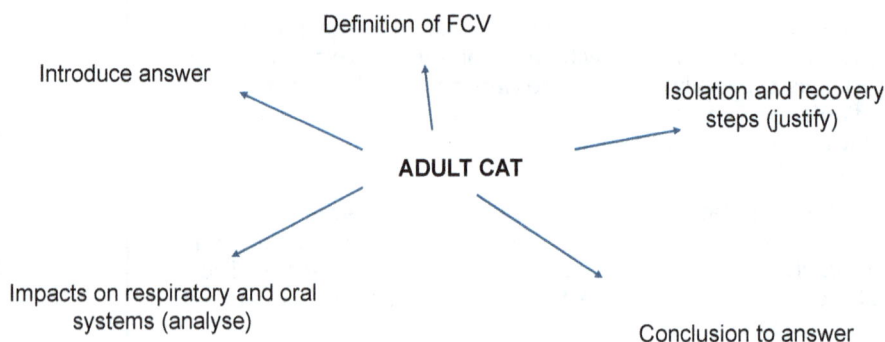

Definition of FCV

Introduce answer

Isolation and recovery steps (justify)

ADULT CAT

Impacts on respiratory and oral systems (analyse)

Conclusion to answer

Template 2: bullet points

- Introduction
  - » Briefly define feline calicivirus (FCV).
  - »
  - »
  - »
- Effects of FCV (analyse respiratory & oral)
  - »
  - »
  - »
  - »
  - »

- Isolation and Recovery steps (justify to ensure health)
  - »
  - »
  - »
- Conclusion
  - »
  - »
  - »

Practice extended response

Answering an extended response question to maximise the marks is as much a skill as knowing the content in the first place. You should practise as many sample questions as you can to help understand the process.

Let's look at the following example 12 mark question:

"A hurricane is predicted to make landfall in the next three days, bringing the risk of extreme winds and heavy rainfall to a wildlife rehabilitation centre and the surrounding area. Due to expected flooding and road blockages, access to the centre may be severely restricted for staff, supply deliveries, and emergency services. Additionally, power outages are likely, and no safe transport is available to evacuate the animals. As a result, a small team of staff will be required to stay onsite during the storm to ensure the animals' welfare.

Evaluate the potential hazards faced by the staff remaining onsite during the hurricane. Propose and justify measures to safeguard both the staff and the animals throughout the period of restricted access."

> **Remember!**
> Practicing exam questions is an important part of revision. It can be difficult to know how to answer a question, or even what it is actually asking!

Pay attention to the command words in the question – they have been highlighted here to draw your attention to them. (Note that these highlights will not be provided for you in the real exam, though you can highlight them yourselves). You must follow the command words or risk not achieving marks in the higher bandings. Very often examiners' feedback from questions like these states that the learners were descriptive only and did not justify and link their answers to the scenario.

Now have a go at answering the broken-down questions below. This is a suggested structure to answer the above question. However, there are many other ways a good answer can be written. Have a go at the other samples afterwards.

Introduction
- Write a short (2-3 sentence) summary of the scenario set out in the question.
- What are the 2 most important considerations in this situation?

Hazards
- Outline the immediate hazards in this scenario (refer to hazards and risks)
- Outline the risks associated with those hazards
- Describe the challenges that the facility could face and how these can be overcome

Recommend Measures for Staff and Donkey Care
- Outline the steps required to ensure staff and animal safety, justifying why you would recommend each step.
- Identify the main barriers to those steps and, again, justify your answers.

Post-Storm Recovery Plan
- How can you ensure safety for staff and animals after the outlined scenario?

Conclusion
- Briefly outline the steps taken and the importance of each, linking back to the scenario and animal welfare.

How many marks would you expect to achieve from your answer? Have you checked if it answers the question? Have you demonstrated your knowledge from across the course?

Here is another example question for you to have a go at, this time without a template structure.

"A rescued adult cat has recently been diagnosed with the bacterial infection, salmonella, leading to significant weight loss. The cat is experiencing vomiting, diarrhoea, and lethargy. Veterinary care and treatment are currently in progress.

Analyse the impact of salmonella on the cat's digestive system. Justify the necessary actions the rescue centre must take to support the cat's health while it remains in isolation and undergoes treatment for the infection. You are not required to discuss the specific veterinary medication prescribed."

Answer in the space provided below and consider if the template style above was helpful or not compared to this. It will help you decide your best approach to answer this type of question.

Revision technique and preparing for exams

General advice

- Ensure all your lecture notes are up to date.

- Check the syllabus for your course to ensure you have material that covers everything on the syllabus for your exam.

- Ensure you understand what will be covered in your exam. Not all units will be in the exam, so check you know which ones are covered.

- Ensure you understand the format of the exam.

- Have a look at how many marks there are in the exam in total, and the length of the exam, to work out an estimate of how much time you should spend per mark during the exam. Do give yourself buffer times at the start and end of the exam.

- For example, a 2 hour (120 mins) exam has 60 marks in total. Giving yourself 10 mins at the start and at the end (20 mins in total) leaves 100 mins to cover 60 marks. This is 1.6 mins per mark approximately. A 12-mark question would therefore be allocated 20min for this example.

- Look at exemplar papers / past papers on the awarding body's website.

- The awarding body's website may also have other resources to help you prepare for the exams.

- Familiarise yourself with the meaning of command words used in exam questions (see section to follow).

- Allow yourself time to revise – work out a plan that avoids only doing last-minute revision. A revision planner needs to work for you and needs to be adaptable as you go along.

- The more times you revisit the topics of the exam, the greater the chance of the information being retained.

- You need to find revision techniques that work for you, and can be a combination of techniques. See the following section for ideas.

- Get plenty of sleep – this is fundamental to good revision practise.

- Reduce stress – there are lots of techniques to reduce exam stress, find ones that work for you.

Actual exam technique

- **Read the whole paper** before you start answering questions – often the first time you read a question you will see what you want to see, not what the question is actually asking. By reading it a second time you are more likely to see exactly what it is asking.

- **Annotate the questions** so you highlight the command words, numbers of examples required, species or named animal required, and any other key detail

- **Look at the marks for each question** so you can decide how much time to spend on each question – it is not advisable to spend a long time on a low mark question

- If your paper is in a booklet format you can **answer the questions in any order** – decide whether to start with the easiest, the hardest or your favourite subject. Just choose the order that works for you, to keep you confident and progressing through the exam. But make sure you don't run out of time for harder questions that are worth the most marks.

- **Pay attention to the instructions on the exam paper**, especially in terms of the pen type you can use, whether you can write in the margins or not (normally you can't) and whether crossed out work will be marked.

- If the exam is in a booklet with lined spaces for your answers and you cannot write in the margins or between questions, then **use extra paper if you need** more lines (as long as this is allowed).

- **Keep an eye on the time** to ensure you have time to answer longer questions worth more marks. If you have a tendency to run out of time in exams, you might want to consider tackling longer questions in the middle of the time allotted rather than leaving them towards the end.

- **Answer every question**, even if you are not sure of the answer, as you may gain some marks that might make a difference between grade boundaries

- If you are unsure of which of two answers are correct, then put both in, in case examiners are allowed to ignore the wrong answer and mark the correct answer

- **Read through your answers at the end of the exam**, check they answer the question and make sense and that you can you see where the marks would be awarded.

- **Halfway through a long answer, pause and re-read the question**. This will help ensure you haven't drifted away from what the question is asking.

Understanding command words in exam questions

Below is a list of common command words used in exam questions, with general explanations and examples to clarify what each term is asking for. Recognising these command words helps you understand how to respond effectively.

Identify

- What it wants: State specific information without extra details.

- Example: "Identify two signs of poor animal welfare."

- Response: Give two clear signs, e.g., "weight loss" and "excessive vocalisation."

- Pay attention to: Marks available - this should guide you with how much to give. In the example above, it should be a 2 mark question, one mark for each sign

Describe

- What it wants: Provide detailed information about characteristics or steps, without explaining why.

- Example: "Describe the process for cleaning an animal enclosure."

- Response: "Remove all waste material, wash surfaces with disinfectant, rinse with water, allow to dry and replace bedding."

- Pay attention to: Marks available, as the more marks the more detail is required.

Explain

- What it wants: Provide details with reasons, showing cause and effect.

- Example: "Explain why regular veterinary check-ups are important for pets."

- Response: "Regular check-ups can catch health issues early, which prevents severe illness and helps maintain animal welfare."

- Pay attention to: Ensuring you are explaining and not just stating or describing, and give examples to illustrate your points.

Outline

- What it wants: Give a general summary or list the main points.

- Example: "Outline the steps to safely handle a nervous dog."

- Response: "Approach slowly, use a calm voice, avoid direct eye contact, and consider using a leash."

- Pay attention to: It is wise to still explain each item.

Discuss

- What it wants: Present a balanced argument, considering multiple points of view.

- Example: "Discuss the ethical considerations of keeping animals in zoos."

- Response: Include points on conservation benefits, ethical treatment, and potential drawbacks, like restricted movement.

- Pay attention to: This should be a detailed answer with 'what', 'why' and 'how' it links.

Evaluate

- What it wants: Assess the strengths and weaknesses, giving an overall judgement.

- Example: "Evaluate the effectiveness of environmental enrichment for captive animals."

- Response: You should discuss the benefits (e.g. improved mental stimulation) and limitations (e.g. lack of natural habitats) before concluding.

- Pay attention to: Give an overall judgment if asked for one.

Compare

- What it wants: Show similarities and differences between two or more items.

- Example: "Compare the handling techniques used for dogs versus cats in veterinary practice."

- Response: "While dogs may respond to direct restraint, cats often require a gentler approach, using towels or minimal restraint."

- Pay attention to: Must give both sides of the comparison to get the most marks

Contrast

- What it wants: Focus only on the differences between two or more items.

- Example: "Contrast the nutritional needs of carnivores and herbivores."

- Response: "Carnivores require high protein, while herbivores need a fibre-rich diet with varied plants."

- Pay attention to: It is still wise to explain why with your contrasting.

Exam and revision tips

Analyse

- What it wants: Break down the topic into components and examine each part in detail.
- Example: "Analyse the factors influencing animal behaviour in a shelter environment."
- Response: Describe each factor, such as confinement stress, interaction with humans, and noise levels, and how each affects behaviour.
- Pay attention to: Giving detailed examples to illustrate your points.

Justify

- What it wants: Provide reasons or evidence supporting a choice or opinion.
- Example: "Justify the use of specific training methods for obedience in dogs."
- Response: "Positive reinforcement is effective because it encourages desired behaviour, increases trust, and reduces fear responses."
- Pay attention to: This command word is used in higher mark questions, so ensure you make enough points with justification and linking to the scenario to equate to the marks.

Summarise

- What it wants: Condense information into brief main points without details.
- Example: "Summarise the main responsibilities of an animal care technician."
- Response: "Ensuring animal welfare, maintaining clean enclosures, monitoring health, and assisting with medical treatments."
- Pay attention to: The marks should help you decide how much detail is required.

Define

- What it wants: Give the exact meaning of a term or concept.
- Example: "Define biosecurity."
- Response: "Biosecurity is a set of measures aimed at preventing the spread of infectious diseases among animals."
- Pay attention to: Number of marks as they dictate how much detail is required.

Revision techniques

Time management

- Revising effectively requires maintaining focus and mental energy, which is why taking regular breaks is essential.
- Breaks help prevent burnout, aid in memory retention, and keep motivation high.
- Try following the Pomodoro Technique, where you study in focused 25-minute intervals with a 5-minute break, or take a 15-30 minute break every hour.
- During breaks, engage in relaxing activities, like stretching, taking a short walk, or practising deep breathing exercises. This helps clear your mind, enabling you to return to studying refreshed and ready to absorb more information.

Blind Mind Mapping

Blind mind mapping is an excellent technique for testing your knowledge of a topic without referring to notes initially.

Start by writing a central idea — such as "Animal Welfare Legislation" — on the centre of a page. Without using your notes, branch out from the centre, adding subtopics like "Acts and Regulations," "Ethical Considerations," and "Impact on Animal Care." Challenge yourself to recall specific legislation, procedures, and ethical issues related to each branch.

After creating the initial map, compare it with your notes. Identify gaps or inaccuracies and correct them on the map in a different colour. This process strengthens memory by highlighting areas you may need to revise further, and it gives you a clear visual representation of how different topics are connected.

Alternative: Use your notes to create the best possible mind map you can, including everything from your notes. Then cover it up and try to recreate it from memory. Challenge yourself by limiting how many times you look at the original for help. Over time, allow yourself fewer and fewer looks at the original.

Flashcards

Flashcards are ideal for memorising definitions, lists, and processes. For Animal Care and Management, use flashcards to remember classifications, behavioural concepts, or care protocols. For instance, create a flashcard with "Biosecurity Protocols" on one side and list steps like "Isolation of New Arrivals" and "Sanitation" on the other.

When using flashcards, employ active recall and spaced repetition. Active recall involves testing yourself on each card, avoiding the urge to look at the answer immediately. **Spaced repetition** means reviewing the cards at intervals, such as daily, then every few days, to help transfer knowledge to long-term memory.

Remember: The key is not to overload a single flashcard with too much information. They should be quick facts. You may even like to write a question on one side with the answer on the back.

Remember: Using spaced repetition of flash cards helps you to identify information which you are best at, as well as information that needs more time to remember. You can then focus your time more on these areas!

Structured Note-Taking

Structured note-taking helps organise information logically, making it easier to study. Use methods like the **Cornell Note-Taking System**, which divides the page into sections for main notes, key points, and summaries. For example, when studying "Animal Handling Techniques", list main techniques in the notes section, jot down keywords like "restraint," "muzzling," and "calming techniques" in the left margin, and then write a summary of handling principles at the bottom.

Organised notes support understanding by clarifying complex topics and facilitating quick review before exams. Use headings, bullet points, and underlining for emphasis, and try colour-coding to distinguish different aspects of a topic, like safety protocols vs. ethical considerations.

Alternative: Use the margin to write questions based on the content of the main body of text it is next to. Then cover up the notes and test yourself on the questions in the margins. For example, when studying "Animal Handling Techniques," the question could be "explain how muzzling can be used as a training aid."

Mnemonics

Mnemonics are memory aids that make complex information easier to remember by associating it with familiar words, patterns, or imagery. In Animal Care and Management, mnemonics can simplify the memorisation of specific processes or classifications.

For example, to remember the sequence for approaching and restraining an animal, create a mnemonic using the first letter of each step, like "CALM" (Check for signs of distress, Approach slowly, listen to cues, Manage movement carefully).

Memory Dump

A **memory dump** is a useful revision technique that involves writing down everything you remember about a topic before checking your notes. Choose a topic, such as "Health and Safety in Animal Facilities," and write down everything you can recall, from safety protocols to common hazards, without looking at your notes.

Once finished, compare your memory dump to your study material, marking any missed points or errors. This activity gives insight into what you already know and where you need more review. By repeating this technique, you reinforce memory and improve recall, especially for exam settings.

Funnelling

Funnelling is a method where you gradually narrow down broad topics into focused details. Begin with a high-level concept, such as "Animal Health and Wellbeing." Start by writing general ideas, like "Indicators of Health," "Diet and Nutrition," and "Behavioural Health." Then, funnel these down to specific subtopics, such as common dietary needs, specific signs of health issues, and behavioural assessment methods.

This technique is valuable for subjects with complex layers, helping you progressively deepen your understanding without feeling overwhelmed. After creating a funnel, practice explaining each layer in your own words to enhance understanding and memory retention.

Explaining the topic to another person

Explaining the topic to someone else can help you retain information and key concepts. Find willing volunteers to listen to you and ask questions. You can also give them questions to ask you to help draw information out.

There are lots of websites and organisations which offer advice on exam techniques, revision methods, and reducing exam stress.

Do some exploring and experimenting and find things that work for you and help you achieve success.

9 781917 048019